In the late 1970s Bernard Stiegler was arrested for armed robbery and imprisoned. Whilst on hunger strike he was given his own cell where, in solitude, he began to study philosophy until his release in 1983. By 1993, under the supervision of Jacques Derrida, he completed his PhD, which was published a year later as Volume 1 of the *Technics and Time* series. Stiegler went on to become one of the most influential philosophers of the twenty-first century.

Stiegler for Architects is the first introduction to the key concepts and ideas of Bernard Stiegler that are relevant to architects. The book asks to what extent it might be possible to have a right to the city in our age of contemporary algorithmic technology. The book begins with a hypothesis: The philosophy of Bernard Stiegler provides an adequate methodology by which we might understand the effects of contemporary digital technology. Second, the fundamental basis of Stiegler's philosophy is introduced—human evolution is not possible apart from technology. Third, the book introduces how his work might be used to think about the city in our contemporary technological age.

The book concludes that the question of the extent to which the right to the city is possible in our contemporary technological age is a question of the extent to which it is possible to prescribe a therapeutics that is capable of being a cure— one that acts across the multiple scales upon which algorithmic technologies operate. This book is essential reading for any architect or designer who is interested in how contemporary digital technology affects everyday life in the city, or anyone wrestling with Stiegler's ideas.

David Capener is an architect, artist, and writer. In collaboration with the multi-disciplinary design group ANNEX he co-curated and designed *Entanglement*, the Irish Pavilion at the *Venice Biennale 2021*. His practice critically examines the spaces produced by our contemporary technological condition. He is co-editor of the book *States of Entanglement* (2021).

Thinkers for Architects

Series Editor: Adam Sharr, Newcastle University, UK

Editorial Board
Jonathan A. Hale, University of Nottingham, UK
Hilde Heynen, KU Leuven, Netherlands
David Leatherbarrow, University of Pennsylvania, USA

Architects have often looked to philosophers and theorists from beyond the discipline for design inspiration or in search of a critical framework for practice. This original series offers quick, clear introductions to key thinkers who have written about architecture and whose work can yield insights for designers.

"Each unintimidatingly slim book makes sense of the subjects' complex theories."
Building Design

"... a valuable addition to any studio space or computer lab."
Architectural Record

"... a creditable attempt to present their subjects in a useful way."
Architectural Review

Kant for Architects
Diane Morgan

Freud for Architects
John Abell

Peirce for Architects
Richard Coyne

Latour for Architects
Albena Yaneva

Baudrillard for Architects
Francesco Proto

Stiegler for Architects
David Capener

For more information about this series, please visit: https://www.routledge.com/Thinkers-for-Architects/book-series/THINKARCH

Stiegler
for
Architects

David Capener

LONDON AND NEW YORK

First published 2025
by Routledge
4 Park Square, Milton Park, Abingdon, Oxon OX14 4RN

and by Routledge
605 Third Avenue, New York, NY 10158

Routledge is an imprint of the Taylor & Francis Group, an informa business

© 2025 David Capener

The right of David Capener to be identified as author of this work has been asserted in accordance with sections 77 and 78 of the Copyright, Designs and Patents Act 1988.

All rights reserved. No part of this book may be reprinted or reproduced or utilized in any form or by any electronic, mechanical, or other means, now known or hereafter invented, including photocopying and recording, or in any information storage or retrieval system, without permission in writing from the publishers.

Trademark notice: Product or corporate names may be trademarks or registered trademarks, and are used only for identification and explanation without intent to infringe.

British Library Cataloguing-in-Publication Data
A catalogue record for this book is available from the British Library

Library of Congress Cataloging-in-Publication Data
Names: Capener, David, author.
Title: Stiegler for architects / David Capener.
Description: Abingdon, Oxon : Routledge, [2025] | Series: Thinkers for architects | Includes bibliographical references and index.
Identifiers: LCCN 2024012884 (print) | LCCN 2024012885 (ebook) |
ISBN 9781032506500 (hardback) | ISBN 9781032506494 (paperback) |
ISBN 9781003399445 (ebook)
Subjects: LCSH: Stiegler, Bernard. | Architecture—Philosophy.
Classification: LCC B2430.S7524 C37 2025 (print) | LCC B2430.S7524 (ebook) |
DDC 720.1—dc23/eng/20240702
LC record available at https://lccn.loc.gov/2024012884
LC ebook record available at https://lccn.loc.gov/2024012885

ISBN: 9781032506500 (hbk)
ISBN: 9781032506494 (pbk)
ISBN: 9781003399445 (ebk)

DOI: 10.4324/9781003399445

Typeset in Frutiger
by codeMantra

For A.

Contents

Series Editor's Preface	viii
Preface	x
1. Introduction	1
2. The Invention of the Human	12
3. Stiegler and the City	30
4. Technology and Memory	42
5. The Poison and the Cure	54
6. Algorithmic Memory	69
References	81
Further Reading	87
Index	89

Series Editor's Preface

Adam Sharr

Architects have frequently looked to thinkers in philosophy and theory for design ideas, or in search of a critical framework for practice. Yet architects and students of architecture can struggle to navigate thinkers' writings. It can be daunting to approach original texts with little appreciation of their contexts. And existing introductions seldom explore a thinker's architectural material in detail. So this original series offers clear, quick, and accurate introductions to key thinkers who have written about architecture. Each book summarizes what a thinker has to offer for architects. It locates their architectural thinking in their wider body of work, introduces significant books and essays, helps decode terms, and provides quick reference for further reading. If you find philosophical and theoretical writing about architecture difficult, or just don't know where to begin, this series is indispensable.

Books in the *Thinkers for Architects* series come out of architecture. They pursue architectural modes of understanding, aiming to introduce a thinker to an architectural audience. Each thinker has a unique and distinctive ethos, and the structure of each book derives from the character at its focus. The thinkers explored are prodigious writers so any short introduction can only address a fraction of their work. Each author—an architect or architectural critic—has focused on a selection of a thinker's writings which they judge most relevant to designers and interpreters of architecture. These books will therefore be the first point of reference, rather than the last word, about a particular thinker for architects. It is hoped that the books will encourage you to read further, inviting you to delve deeper into the writings of a particular thinker.

The *Thinkers for Architects* series is now in its second decade. Books published so far have mostly covered so-called canonical figures: Well-established names from the traditions—the 'canons'—of philosophy, and critical and cultural theory, who have long influenced architecture and architects. Like in most academic and professional fields, such names have historically been largely Western, white and male. This reflects the structures of power and influence that typically determine who can access education, publishing, reviews, and the kinds of jobs which give people space to think and write. We live at a time when populist politics are on the rise in various countries around the globe. Some governments, and certain media organizations, are weaponizing intolerance of differences between people for political ends. Against this context, it grows ever more urgent to appreciate who is—and who isn't—typically able to speak to a public audience, and to diversify the reference points of academic disciplines and professions like architecture. From the start, *Thinkers for Architects* has always been concerned with ideas that challenge traditional canons. Having covered many long-established thinkers, the series now seeks to engage a wider range of voices. Alongside familiar names, it will increasingly introduce thinkers who are less familiar to architects, but whose ideas have equal potential for designers. For example, as well as books covering Georgio Agamben, Hannah Arendt, Graham Harman, and Bernard Stiegler, planned volumes will examine the architectural thinking of Franz Fanon, Stuart Hall, Donna Harraway, McKenzie Wark, and Simone Weil. The series continues to expand, aiming to explore exciting and diverse thinkers who have something to say to architects.

Adam Sharr is Professor of Architecture at Newcastle University, UK, and Editor-in-Chief of *arq: Architectural Research Quarterly* published by Cambridge University Press. He practices with Design Office, which was included in the *Architects Journal* '40 under 40' listing of 'the UK's most exciting emerging architectural talent' in 2020. Adam is author or editor of eight books on architecture, including *Heidegger for Architects* and *Reading Architecture and Culture* published by Routledge.

Preface

This book has come from a longstanding interest in the effects that contemporary digital technologies have on everyday life. The question that has driven my research is a simple one: How as architects can we imagine and realize new ways of being in the world—ways of being that bring justice and hope? I am interested in exploring what methodologies might be adequate to this task. I propose that the work of Bernard Stiegler provides one such useful methodology. This does not mean that there *are* others, my hope is that there are, but through this book I hope to contribute to that conversation which is ultimately a question of the extent to which a right to the city is possible in our age of contemporary digital technology.

CHAPTER 1

Introduction

Our age of contemporary digital technology is a material and spatial reality. Everyday life is increasingly becoming governed by algorithmic technologies that not only produce what we do in space but are producing new kinds of spaces. I argue that this is an architectural problem. I propose that architects should seek out theoretical tools to help understand these spatial transformations. What kinds of spaces are being produced by the shifts in the ways that we understand and navigate everyday life facilitated by our contemporary technological condition? How are these spaces being produced? Why are they being produced? This book argues that architects need adequate methodologies to understand the complex processes of spatial production in our age of contemporary technology. This book introduces some of the fundamental ideas in Stiegler's work. It shows how they are relevant to living in the city in our age of contemporary digital technology. The premise of this book is not to ask you the reader to fit the round peg of Bernard Stiegler's philosophy into the square hole of architectural practice. Stiegler's philosophy does not fit neatly into architectural practice.

<u>The hypothesis of this book is that Stiegler's philosophy asks questions about what it means to practice architecture in an age where everyday life is becoming increasingly transformed by contemporary digital technologies. As we face into an uncertain future it is the square peg that is in question.</u>

Bernard Stiegler

Bernard Stiegler is one of the most influential philosophers of the twenty-first century. His journey to becoming a philosopher was turbulent. After becoming a father at the age of 19 his struggle with alcoholism saw him admitted to a psychiatric hospital from which he escaped. Homeless, he lived on a farm, in a car, with a pet monkey called Zoë. He went on to convert a brothel into a jazz club, which became a local haunt for gangsters selling heroin. He refused to cooperate with the police, so they forced him out of business. Desperately in need of money Stiegler turned to robbing banks. After some success he was caught and sent to prison for five years where he went on hunger strike. This resulted in him being given his own cell where, in solitude, he began to study philosophy until his release in 1983. On his release under the supervision of philosopher Jacques Derrida, he completed his PhD, which was published a year later as Volume 1 of the *Technics and Time* series. Stiegler went on to become one of the most influential philosophers of the twenty-first century.

The first time I met Bernard was in Dublin at the European Artists Research Network conference in 2018. I was presenting a paper titled 'The Right to the City: A Revolutionary Conception of Citizenship' where I was beginning to think through Stiegler's philosophy and the right to the city. Afterwards Bernard came up to me, shook my hand and said 'hello I'm Bernard, I really enjoyed your paper, it seems we share some similar interest, would you please send it to me, and if you would like to come and visit me in Paris we could find ways to collaborate.' Bernard had a generosity of spirit. I never made it to Paris, but I remember the encouragement I felt in that moment, that maybe, in those naïve and wandering ideas there was something worth perusing. On 5 August 2020, aged 68, Stiegler died.

A Note to the Reader

As with the work of many philosophers, to engage with the work of Stiegler is to be invited into a new way of thinking about and seeing the world. This new world has its own often difficult language, populated with strange sounding concepts that can often alienate a reader who is approaching his work for the first time. This book does not avoid using these words. It is important to introduce you the reader to them, not simply because they name key ideas but also because the words that Stiegler chooses to name his concepts have a history that furnishes them with a depth and richness. For example, his concept of pharmakon introduced in Chapter 5 is an idea as old as Western philosophy and tells the story of how philosophers since Plato have wrestled with ideas about the positive and negative effects of technology. While it is important to introduce you the reader to these concepts you are not asked to walk through a dense forest of unfamiliar concepts through which you might become lost. While this book cannot avoid these often-difficult concepts it is my hope that a clear path has been laid along which you can travel. It may be that not everything becomes immediately clear. I myself still wrestle with Stiegler's ideas. If this happens I would encourage you to keep walking. It is my hope that this book will be of interest to anyone encountering Stiegler's work for the first time.

Why Stiegler?

To answer the question of why architects should be concerned with our contemporary technological condition is, I argue, to learn to think differently about architectural practice or rather the future of architectural practice, which is also to think about the present of architectural practice. It is to think beyond the red line boundary of an individual site and consider the entanglement of technological infrastructures that are now integral to everyday life.

Ultimately this is to reframe what we mean when we talk about a site or a client or context, scale, and perspective.[1] It is to ask the question who, in the midst of the climate emergency, is our client? Which is a question of how to pay

---

[1] I am grateful to the Architect and Educator Aoife McGee for the idea that architects increasingly need to think beyond the 'red line' constraints of their site and for helping me recover an understanding of the importance of the role of the architect in the production of the commons in everyday life.

attention to so-called other-than-human clients against which our discipline has performed great injustices and violence. How can we find ways of practicing that become attentive to, and indeed platform, these other voices? How, as architects, might we learn to give space to both human and other-than-human voices—to those who speak in unfamiliar tongues? How might we learn to become translators of these strange, foreign languages? How do we learn to speak the language of a tree? How might we translate the language of an algorithm?

This is a question of redefining how we think about the context of architecture, which involves calling into question human-centric ideologies of scale and perspective. Which context? At what scale are we to think? From whose perspective do we view? These are necessary questions for architects to consider in our age of contemporary algorithmic technology. There is an urgency to this task—if it is not already too late.

Perhaps the question of how we will live together is now, in the midst of a climate catastrophe, really a question of how we will die together? To what extent might we find ways to design worlds of justice, equity and hope in the face of the inevitable.

This book argues that the question of the usefulness of Stiegler's work to the practice of architecture in our age of contemporary algorithmic technology is the question of how the practice of architecture can contribute to the production of what philosopher Henri Lefebvre called the 'right to the city' (Lefebvre, 2008, p. 158). At its core, the right to the city, for Lefebvre, is a 'cry and demand' for 'nothing less than a revolutionary conception of citizenship' (Lefebvre, 2008, p. 158). This conception of citizenship would 'break up systems' and 'open up through thought and action towards possibilities by showing the horizon and the road' (Lefebvre, 2008, p. 158). The right to the city is also not just a material right that one might exercise over a geographic locale, a neighbourhood or a building, or a right to the basic necessities such as water, food, shelter, or education as David Harvey foregrounds in his popular essay on the right to the city (Harvey, 2003, p. 23–40). The right to the city is, as Lawyer and Planner Peter Marcuse describes it, 'a theoretically complex and provocative formulation' (Marcuse, 2009, p. 185–197). My argument is as follows: The right to the city is something that must be enacted. It is the right to produce spaces, not spaces of capital—what Lefebrve calls homogeneous 'abstract space,' the space of capitalism—but spaces of justice and equity (Lefebvre, 1991, p. 33, 49–53, 60). These spaces counter the homogenous nature of abstract space. They are what Lefebvre calls spaces of difference. Ultimately the right to the city is 'the right to difference' (Lefebvre, 2009, p. 120). This was also a concern of Stiegler's.

Stiegler and the Right to the City

It could be argued that the right to the city was a central concern for Stiegler who founded The Real Smart Cities project with collaborators Professor Noel Fitzpatrick and Dr Gerald Moore. The project investigates the ways in which so-called 'smart city' technologies present a threat to democracy and citizenship, and how digital tools might be used to create new forms of community participation.[2] The philosophical underpinning of the project is that technologies have the potential to be both toxic and curative. Stiegler calls this 'pharmacology', a concept that I will explore more fully in Chapter 5. One project that looks to solve some of the toxic problems produced by contemporary technologies is what Stiegler calls a clinic of contribution, within Pleine Commune in greater Paris (an area where one in three live in poverty). The clinic attempts to deal with issues of communication between parents and children where the widespread use of smartphones from infancy is having effects on the attention of young children and on communicative abilities between parents and children. This forms part of a wider project in that area, which Stiegler describes as 'installing a true urban intelligence.' This moves beyond what he sees as the bankrupt idea of smart cities—often the implementation of

---

[2]A version of this section appeared in David Capener, 'In Many Ways Smart Cities are Really Very Dumb.' www.citymetric.com/fabric/many-ways-smart-cities-are-really-very-dumb-4384

technologies of governance deployed to control how people inhabit the city for reasons of increased productivity. The aim is to create a 'contributory income' in the area that responds to the loss of salaried jobs due to automation and the growth and spread of digitization. The idea is that an income could be paid to residents, on the condition that they perform a service to society. This, if you are unemployed, living in poverty and urban deprivation, sounds like a simple and smart idea to try and solve some of what Stiegler would call the stupid effects of the digital technologies that are being implemented in cities under the ideology of being smart. This questions the extent to which any kind of a right to the city is possible in our age of contemporary digital technology. The right to the city in our age of contemporary algorithmic technology is a design problem. My book argues that this design problem requires design methods, both theoretical and practical, that are capable of understanding and acting upon the multiple scales—from planetary cable networks, to regional data centres, to devices the size of our pocket, to the desires they produce and behaviours they effect—across which contemporary algorithmic technologies operate. It is the challenge to find a 'response adequate to all the systemic challenges arising in the face of contemporary concrescence'—concrescence being the complex entanglement of scales across which contemporary technology operates (Stiegler, 2018a, p. 31).

At the basis of Stiegler's philosophy is a foundational idea: Reality has always been augmented, and intelligence is always artificial. From prehistoric cave paintings—pigment blown onto hands; silhouettes on a cave wall; a series of 9000-year-old selfies—to the estimated 92 million selfies uploaded each day to Instagram, the story of augmented reality and artificial intelligence is not new (Broz, 2024). The history of augmented reality and artificial intelligence is the story of how we became human.

Indeed it would not have been possible to become human apart from the technologies that we formed, and that formed us. Stiegler argues that the human is an invention, and the evolution of the human would not have been possible apart from the co-constitutive relationship that we have with technology. By co-constitutive, Stiegler means that technology is not simply instrumental, something that sits outside of human evolution, but is part of what it means to become human. It is not possible to understand Stiegler's philosophy without understanding this relationship. This is fully introduced in Chapter 2.

Chapter 2—The Invention of the Human

In Chapter 2 the relationship between human evolution and technology is introduced through a question that is fundamental to Stiegler's philosophy. Is technology simply instrumental—which is to say a tool by which we act upon the world? Or, to what extent does technology act upon us? Chapter 2 explores Stiegler's answer to these questions by introducing two concepts from which any journey through his work must begin. These starting points are the concepts of endosomatization and exosomatization. The endosomatic is that which belongs to an individual organism from birth—a limb, wing, eye, or finger. The exosomatic are technologies produced by an individual that were not part of that individual from birth. To explain the relationship between these two concepts philosopher Maurice Merleau-Ponty's example of a blind person's cane is introduced. The cane is what Stiegler would call an exosomatic technology. But, Stiegler poses a question: Is a visually impaired person's cane just a white stick that acts as tool for getting from one place to another? Is it just instrumental? Or is the stick something more? Bernard Stiegler would argue yes, the stick is something more, it is a kind of extension of the body, a technical prosthesis. It is not simply a prosthesis that stands apart from the body, but one that is intimately intertwined with the functioning of the body's apparatus of perception.

The stick supplements the reduced perception of the user, alongside their biological organs, the eyes, hands, ears. The non-biological organ of the stick becomes just as much part of their perceptual apparatus. The world cannot be perceived and known apart from the stick—the way it feels, how the body's nervous system responds to the rumble of the vibrations as it moves across different surfaces, how new neurological pathways are formed as the brain learns the tiny differences between sensations, and the difference between the pavement, edge of the kerb and the road. The stick becomes a non-biological technical organ connected to the biological, organic organs of the body. As the perceptual apparatus of the person evolves it does so because of the use of the stick. The stick is a kind of memory device that exists outside of the body—a third kind of memory. Stiegler calls this third memory tertiary retention.

Chapter 3—Stiegler and the City

Chapter 3 builds upon the idea of the co-constitutive nature of technology and human evolution (or the endo and exosomatic) introduced in Chapter 2 by investigating how Stiegler's work might be useful for thinking about the city in an age where contemporary algorithmic technology increasingly complicates the circuit of subjectivity. By subjectivity I mean the process by which we become subjects not simply as thinking beings but as beings that experience the world through our whole bodies.

I argue that the way in which the infrastructures of our contemporary technological condition transverse multiple scales creates evermore entangled circuits of subjectivity between human and non-human cognitive agents

(Berry, 2018, p. 40). I take the idea of non-human cognitive agents from N. Katherine Hayles who outlines a difference between agents that have both consciousness and cognition (humans) and agents that are non-conscious but cognitive in that they can make some kind of sense of and act in the world (a facial recognition algorithm) (Hayles, 2017, p. 2). To explain this new relationship, and to help think through its implications for the city, the concept of 'algorithmic governmentality' is introduced (Rouvroy and Stiegler, 2016). Algorithmic governmentality was first introduced by Antoinette Rouvroy to describe the governance of the world based on the algorithmic processing of big data (Rouvroy, 2020). To help understand how algorithmic governance is increasingly entangling the endo and exosomatic in more complex ways a further development of these concepts is introduced—infrasomatization (Berry, 2016). In his later work Stiegler uses the concept of infrasomatization to describe how algorithmic technologies are creating new cognitive entanglements between technological infrastructures and the body. Drawing from the Latin, infra means 'below' 'or under a part of the body' and somatic means the body (Berry, 2016). So put simply the concept of infrasomatization tries to capture the idea of technological infrastructures that operate outside of the normal range of perception yet are intrinsically entangled with the body. To develop this idea and show how it might be used to think about the city in our age of algorithmic governmentality the concept of exorganic milieu is also introduced. For Stiegler an exorganic milieu is composed of infrastructures that operate simultaneously across multiple scales from the micro-biological to the planetary.

Chapter 4—Technology and Memory

Chapter 4 explores Stiegler's understanding of memory by introducing a three-fold categorization of what he calls retention: Primary, secondary and tertiary retention. Continuing with the example of the blind person's stick, primary retention is introduced as the visually impaired person's direct perception of the world, their primary memory. It describes their direct perception of the sounds and smells as they walk through the city. Secondary retention comprises stored memories—certain places that the person has been before, including past sounds and smells. The stick then becomes a third kind of memory, a memory that exists outside of the human body, what Stiegler calls a tertiary memory device (Stiegler, 2011, p. 26). Without the tertiary memory device of the stick, and these three types of memory working together, the visually impaired person would experience the city very differently. But what happens when the loop between the technical prosthesis and the body becomes complicated by algorithmic technologies, especially when they operate at such a speed that they are able to bypass normal human cognition? Algorithmic technologies that operate 'prior to the 100-millisecond range where perception and sensation register for humans' (26). What methods and methodologies are capable of understanding the effects that these kinds of technologies have? This will be the subject of Chapter 6.

Chapter 5—The Poison and the Cure

Chapter 5 introduces a further concept central to the work of Stiegler: Pharmakon. Pharmacology and organology are two concepts through which Stiegler tries to think technics (Stiegler, 2011, p. 9). For Stiegler technics means more than just technology in the sense of an instrumental tool. Technics means technology in a much broader sense—the exteriorization of memory through technological prosthesis. Through his concept of pharmacology Stiegler wants to explore how a technology might be both toxic and curative, and how the therapeutic potential of a given technology might be prescribed. This question is explored alongside organology. This is because any potential therapeutics cannot be found in technology alone but must be thought across the biological, psychic, and social organs not just of the individual but of society. Ultimately the question of pharmacology is a question of hope. It is a question of how technology might become the cure rather than a poison. It is a question of the extent to which in our co-evolution with technical supplements we might find a right to the city. This is significant to architects because contemporary technological systems are increasingly entangled with everyday life.

Chapter 6—Algorithmic Memory

Chapter 6 further explores Stiegler's concepts of pharmakon, organology, and neganthropology, understanding them as the basis of working towards a right to the city. The question of how the concept of pharmakon is able to account for the changes brought about by our contemporary algorithmic age is considered. A question is posed: What might a pharmacology of technologies that operate below the threshold of human cognition look like? The chapter argues that as memory becomes delegated to contemporary algorithmic technologies the extent to which a right to the city is possible is curtailed because our perception of space is increasingly becoming mediated below the threshold of human consciousness. To understand these changes an advancement is made to Stiegler's framework of retention. The idea of a primary retention that is non-conscious is introduced.

CHAPTER 2

The Invention of the Human

Introduction

Over the past decade contemporary digital technologies in the form of mobile devices such as smartphones and wearables have become embedded into everyday life. Everyday life is becoming increasingly mediated through these technologies. To navigate life without them no longer seems possible—this is the water in which we now swim. Yet, these technologies are not simply a means by which we navigate the world, they are also navigating us. Through various forms of data mining, our desires, thoughts, impulses, preferences, perceptions, and much more have become valuable commodities. Consumer anxiety produced by the perpetual cycle of upgrades and updates perpetuate the myth that we need the product when in fact it is the product that needs us.

In our age of contemporary digital technologies, we are as much a node in the infrastructure of networked communication as a cable, code, or server. Our age is one of complex technological entanglements. Entanglements that move across multiple scales from the microscopic neurological flows to urban and planetary scale infrastructures. What's more these technologies are now operating at speeds often below the level of human perception.

Bernard Stieger argues that the unprecedented speed at which these technologies can operate, coupled with the levels of access that they now have to our perceptual apparatus, means that both time and space are being completely reconfigured (Stiegler, 2011, p. 120). Using the work of Stiegler, in the rest of this book I explore what some of these spatial transformations might be and how architects and other associated disciplines can find adequate methodologies to understand them. The argument of this book is that Stiegler's philosophy of technology offers one such useful methodology. This methodology begins with an idea that is the foundation of all of Stiegler's philosophy: To be human is to be a technological invention. Any journey through the work of Stiegler must start here with 'the invention of the human' (Stiegler, 1998, p. 127). This chapter introduces this foundational concept.

The Invention of the Human Is Technics

For Stiegler the human is an invention of technology. The human does not just invent technology, but the human is itself an invention of technology. As he puts it, the invention of the human is technics (127). Indeed, without technology, there could be no evolution of the human. This idea leads Stiegler to conclude that when we think about human bodies we must not only think about biological organs, but we must also consider technical organs.

The human does not exist apart from both biological and technical organs. The co-constitutive biological and technical organs of human evolution must always be thought in the context of society. For Stiegler, society comprises four organs, the technical, biological, psychic, and social. These organs are ever evolving, and for Stieger any methodology adequate for understanding our technological age must be able to understand the evolving nature of these organs and both the positive and negative effect that technology has. Stiegler calls this methodology 'organology,'

and the toxic and curative nature of our technological age 'pharmakon.'

These concepts are discussed more fully in Chapter 5, however it is important to note that, for Stiegler, technology has the potential to be both toxic and curative. If the human is an invention of technology then the question for Stiegler that follows concerns how a given technology is either toxic or curative. Discerning between the two Stiegler calls pharmacology. The concept of organology means that this question cannot be considered apart from an analysis of the organs of society. This then leads to a question of pharmacology—the prescription of any therapeutics cannot be found in technology alone but must be thought in context, across the biological, psychic, and social organs of society. This is because technology cannot be thought apart from its co-constitutive relationship with the human. This question cannot be thought apart from the co-constitutive nature of the evolution of the human and technology. The next section introduces this foundational idea in more detail.

The Evolution of Technological Prostheses

Stiegler describes the central concept of the technical evolution of the human in the following way:

> The evolution of the 'prosthesis,' not itself living, by which the human is nonetheless defined as a living being, constitutes the reality of the human's evolution, as if, with it, the history of life were to continue by means other than life. (Stiegler, 1998, p. 50)

What does Stiegler mean by this? Put simply technological prostheses are tools and technologies that are external to the human body. For example, a smartphone is not physically part of the human body (although it does physically affect the human body, for example algorithms that produce certain behaviours and addictions), nor is it a living organism. However, while the smartphone is a technology external to the human body it is not simply a tool in the instrumental sense by which we do things in the world like take photographs, send text messages, or access the internet.

The smartphone is not just something that we use to act upon the world, but is something that acts upon us. It is part of our evolution without which, according to Stiegler, human evolution would not be possible. This happens to such an extent that, for Stiegler, life cannot continue apart from technologies that are external to the human body—be they prehistoric flint used to carve and cut, a hammer and a nail, or a smartphone.

For Stiegler the invention of the human has been forgotten by Western philosophy. A schism has been created between *tékhnē* (technics) and *epistēmē* (knowledge). Over the three volumes of his foundational *Technics and Time* series, Stiegler seeks to bridge this chasm.

A helpful example is the blind person's cane, as explained by the philosopher Maurice Merleau-Ponty. Merleau-Ponty tries to bridge the chasm between technology and the formation of human subjectivity, and to describe the co-constitutive relationship between technical and biological organs with the example of a blind person's cane. The blind person's cane is not simply a piece of technology that is used as an external handheld device to navigate the world. The cane becomes an extension of their perceptual apparatus. Without it, the user's perception of space is significantly curtailed. The cane, the hand, the arm, the nervous system, the brain, all form what can be called a circuit of perception. The cane becomes a technological prosthesis that forms a co-constitutive relationship with the biological organs of the user. The cane is not simple an extension of the subject, a way by which the subject navigates the world, but is constitutive of the subject. The blind person and the cane become entangled. To use Stiegler's terminology the blind person is an invention of the technology of the cane. The human invents the cane, and the cane invents the human.

The relationship between the blind person and the technology of the cane is relatively simple. The cane is an extension of the blind person's perceptual apparatus that exists external to their body. While using the cane produces a circuit of subjectivity that circuit is broken when the blind person stops using the cane. If the blind person tries to navigate space without the cane, their perception of the surrounding environment is entirely different from if they were using the cane. While the cane undoubtedly has a life-changing effect for how the blind person navigates the world, it also has limitations. The circuit of subjectivity is disrupted when they no longer use the cane, but what happens when the possibility exists for this process to be short-circuited?

What happens when the technological prosthesis becomes so integral to the functioning of everyday life that everyday life cannot function without it? And, what happens when it becomes increasingly difficult to separate the technological prosthesis from the body that the components of the 'circuit of subjectivity' are almost indistinguishable from each other? These are questions that Stiegler explores through his concepts of 'endosomatization' and 'exosomatization.'

Endosomatization and Exosomatization

Stiegler develops the concepts of endosomatization and exosomatization to explain the relationship the human body has with technology—like the relationship between the blind person and their cane. Stiegler argues that the human is not just made up of biological organs that are internal to the body, but that we are also made up of technical organs which are external to the body. Biological organs that are internal to the body he calls endosomatic, and technical organs that are external to the body he calls exosomatic organs. This special relationship that the human has with exosomatic organs means that the human is essentially a technical being and to be human means to 'exceed the biological'—put simply to be human is to be more than the sum of our biological organs (Stiegler, 1998, p. 50). The importance for Stiegler of human evolution being more than biological cannot be overstated. Following André Leroi-Gourhan, he argues that the evolution of the human is co-originary with technology—the biological human and technological prostheses cannot be separated. Based on the idea of the co-constitutive nature of the human and using his concepts of organology and pharmacology, Stiegler tries to understand the shifts in technology produced by these biological and technical organs.

Stiegler (1998) develops the concepts of the endosomatic and exosomatic based on the work of mathematician Alfred Lotka (1925), and Nicholas Georgescu-Roegen (1970), and philosopher Karl Popper (1972). Georgescu-Roegen's definition of the two terms is useful and worth quoting in full.

> Apart from a few insignificant exceptions, all species other than man use only endosomatic instruments—as Alfred Lotka proposed to call those instruments (legs, claws, wings, etc.) which belong to the individual organism by birth. Man alone came, in time, to use a club, which does not belong to him by birth, but which extended his endosomatic arm and increased its power. At that point in time, man's evolution transcended the biological limits to include also (and primarily) the evolution of exosomatic instruments, i.e., of instruments produced by man but not belonging to his body. That is why man can now fly in the sky or swim under water even though his body has no wings, no fins, and no gills. (Georgescu-Roegen, 2011, p. 58–92)

Thus, the endosomatic is that which belongs to an individual organism from birth. The exosomatic are technologies produced by an individual that were not part of that individual from birth. Stiegler makes it clear that the exosomatic is not something that the human simply uses, what is only external to the individual, but is in fact constitutive of the human—'the pursuit of the evolution of living by other means than life' (Stiegler, 1998, p. 135). This special relationship that the human has with exosomatic organs means that the human is essentially a technical being and to be human means to 'exceed… the biological' (135). The human is essentially defined by '[t]he evolution of the "prosthesis"' (135). Developing Lotka's concepts further, Stiegler and the Internation Collective (2021) use the term exosomatic to describe the city as 'aggregates and localizations of complex exorganisms within which simple exorganisms cooperate (Stiegler and the Internation Collective, 2021, p. 67). I discuss this more fully in Chapters 4 and 5 when I introduce the way that in our contemporary technological age, the endo and exosomatic become entangled in increasingly complex ways, across multiple scales often operating below the threshold of human consciousness.

The Production of Bodies

The production of bodies by new technological advancements is not new. However, our age of contemporary digital technology produces new relationships between the body and the technological prosthesis.

The advent of the steam engine and the railway produced such shifts by creating the ability to travel at previously unimaginable speeds, through inaccessible areas to places one could have only imagined, which radically changed the traveller's perception of space. The way that space had been perceived was destroyed as localities previously inaccessible, or accessible only by means of cart and horse, collide and 'move into each other's vicinity,' losing 'their old sense of local identity, which used to be determined by the spaces between them' (Schivelbusch, 2014, p. 45).

This spatial collision radically changed the subject's perception of space and time. The idea that technology can change time and space is not new. Neither space nor time is possible apart from the technologies that produce them. Wolfgang Schivelbusch, in his book *The Railway Journey: Trains and Travel in the Nineteenth Century* (2014), shows how perception was radically transformed by technologies of industrialization, in particular the rail network. 'The isolation of localities, which was created by spatial distance, was the very essence of their identity, their self-assured and complacent individuality.' This spatial collision radically changed the subject's perception of space, time, speed, geographies, and other people. Technology thus structures perception and produces new kinds of space.

Contemporary algorithmic technologies have become increasingly embedded into everyday life. For example, real-time digital mapping facilitated by platforms such as Google Maps creates shifts in our perception of the world. As Adam Greenfield writes, when we open something like a digital mapping application 'our sense of the world is subtly conditioned by information that is presented to us for interested reasons, and yet does not disclose that interest' (Greenfield, 2017, p. 63). The information that I see on my screen may be different from information that you see on your screen because (depending on various privacy settings) that information has gone through a process of algorithmic sorting based on my web search history, other apps I have installed, things that I have said, and places that I have visited before. In this way contemporary digital technologies produce new circuits between the body and the infrastructures of everyday life. David Berry (2016) calls this process infrasomatization. The concept of the infrasomatic is a way of foregrounding the ways that contemporary digital technologies produce new circuits between the body and the infrastructures of everyday life (Berry, 2018).

<u>Our perception of space is increasingly becoming algorithmically mediated. This has a tangible effect on our daily routines. According to Google, 'four out of every five consumers use the map application to make local searches' half of those end up</u>

'visiting a store within twenty-four hours, and one out of every five of these searches results in a "conversion," or sale' (Greenfield, 2017, p. 63). These new ways of navigating everyday life brought about by various forms of social media like Twitter, Facebook, and Instagram all produce new ways of being in the world.

What's more, the technologies that govern these transformations are increasingly deployed in the urban realm and move across multiple scales from the microscopic to the body, the urban, and the planetary. At the microscopic scale of neurons and dopamine, desires to visit a certain place or shop in a certain store are becoming increasingly governed by algorithms like those deployed in social media platforms such as Instagram or navigation software like Google Maps. Yet the production of these desires cannot be thought apart from the planetary scale infrastructures like hyperscale data centres that facilitate them. The body and technological infrastructures cannot be thought apart, nor can the rhythms of everyday life that they produce. What do these complex new entanglements do to the concepts of endosomatization and exosomatization? David Berry argues that the complex multi-scalar infrastructures of our contemporary technological condition 'create networks of cognitive agents to commodify human capacity for reason and thought' (Bigo et al., 2019, p. 48). Like Stiegler, what Berry means is that to be human—as a thinking acting being (a cognitive agent)—cannot be considered apart from technology. In our age of contemporary algorithmic technology this relationship has become increasingly more complex. Berry calls this process infrasomatization.

Infrasomatization

I further develop the concept of infrasomatization in Chapter 3, where I will discuss Stiegler's concept of grammartization and introduce the idea of the infrasomatic city. I use the concept of the infrasomatic city as a way of thinking about how contemporary technological must be thought not only in relation to the body as tools that we use, but how they should be thought as constitutive of the body. I argue that not only should we think about the evolution of the human body as co-constitutive with technology but in our contemporary technological age the body must also be thought as an infrastructure in a multi-scalar network of digital technologies.

<u>In our age of contemporary technology, the body has become an essential resource mined by algorithms for its valuable deposits and neurological strata rich in desires for things we never knew we wanted. These multi-scalar entanglements include neurological flows, cables, hyperscale data centres, machine learning models and the human labour involved in the production of image labelling used for training image recognition algorithms.</u>

What the concept of infrasomatization describes is important for understanding the effects of contemporary algorithmic technology—not only has everyday life become colonized by the increasingly pervasive influence of digital technology, but the body itself is becoming colonized. What's more, the body as resource must be thought as not separate from the infrastructures of contemporary digital technology but as infrastructure (Berry, 2018). As such, the body and the infrastructures of everyday life in the city cannot be separated.

In our age of contemporary digital technology, the body becomes a key constitutive infrastructure in the city. Therefore, digital technologies like algorithms must be thought as 'not merely tools' in a Stieglerian exosomatic sense. They must be thought as constitutive infrastructures that produce the very conditions for thought and action by creating new circuits between the body, the brain, and platform infrastructures. The example of the blind person's cane is limited when thinking about algorithmic technologies that are exosomatic. This is because they are external to the human body, but are also endosomatic in the way that they access biological organs often below the threshold of human perception, making interventions on human experience 'prior to the 100-millisecond range where perception and sensation register for humans' (Berry, 2018).

Algorithmic technologies do not just mediate our experience of the world. These technologies are entangling endosomatic organs (human biological organs) together with exosomatic prostheses (algorithmic technologies) in new ways to the extent that the body itself becomes an integral part of the network of technological infrastructures. This new relationship that the body has to technological infrastructures and the access that technology now has to the body means that new kinds of endosomatic evolutions are being produced such as new neurological flows which produce new ecologies of attention and addiction (Citton, 2017).

A useful way to illustrate the complex entanglements between the body and the infrastructures and architectures of our contemporary technological condition is through the example of the processes that take place in online gaming—specifically one-on-one fighting games. This may seem like a strange example but is useful because it extends the example of the blind person's cane and shows how algorithmic technologies produce complex circuits of infrasomatic subjectivity that operate simultaneously at speeds below the threshold of human perception, across multiple scales from the body to planetary scale infrastructures.

Unstable Networks

In one frame of a typical fighting game like *Mortal Kombat*, a lot happens. In the 16 millisecond duration of a frame, a circuit of information is computed. Players are asked for inputs, the network is checked for new information, AI is used to run CPU players, characters are animated, and the current game state is assessed. Who is getting hit? What move have they just made? If latency drops, then this circuit is interrupted, and gameplay is affected. If two players are playing locally then delay is not an issue. However, when two players are playing over the internet then the game needs to account for network instability. Simply put, the information from one player's input will take time to travel through the network and arrive at the other player whose input will take time to travel back. The distance between two players is measured in ping, which is the round trip of information from one player to another and back again. For example, at the time of writing the ping time between Dublin and Boston is 84.35 ms (WonderNetwork, n.d.). This means that a packet of data will travel the 9620 km round trip between Dublin and Boston in 0.08 seconds.

So, for ease let's assume that the Dublin to Boston ping time is 90 ms. This means that, on average, it will take a packet of data 45 ms to travel from Dublin to Boston. Assuming a 16-millisecond frame rate this means that three game frames will have elapsed by the time a packet of data leaves Dublin and lands in Boston. This presents a problem. Central to the gameplay of any fighting game is speed. To play a fighting game well requires muscle memory, twitch reflexes, and very fast player reactions. The instability of networks means that information sent between players may be delayed, or packets might drop or arrive in the wrong order. Network instability is unpredictable in ways that a game cannot predict. For fighting games, the instability of networks is a particular challenge. This game genre cannot function correctly without consistent latency. The question then is how does a game handle uncertainty? This is a question about the volatility of latency. It is also a question of the architectures and infrastructures required to facilitate the passage of data.

Latency

Latency in online gaming can be the difference between winning and losing a match. This is because the threshold of human perception in gaming is much lower than it is for other mediums. For example, when watching online streaming content such as *Netflix*, it is unlikely that a person would notice if the audio was out of sync with the video providing the audio arrives is no more than 45 milliseconds early or 125 milliseconds late. In fact, people will continue to watch the content even if the audio arrives 90 milliseconds early or 185 milliseconds too late. Likewise, on YouTube, only if there is a delay of more than 250 milliseconds between clicking the pause button and the content pausing do we think that something has gone wrong. However, in online gaming, the threshold is significantly lower. Regular gamers will become frustrated with delays over 50 milliseconds, non-gamers at 110 milliseconds, with games becoming unplayable at 150 milliseconds. Indeed, weekly gameplay reduces by up to 6% if there is a 10 millisecond increase in latency. Networks require hardware and software solutions to try and overcome the volatility of latency.

Geopolitics of Latency

One such hardware solution is the use of server regions which aim to minimize latency on a geographical basis (Ball, 2021; Starosielski, 2016). Players are grouped together into geographical regions—such as Western Europe, Midwest America, Southeast Asia. For example, during online play in the car racing game *Mario Kart 8*, players can select to race against other players in their region or globally. When racing globally, latency spikes occur regularly, the game ends and a screen is displayed stating that a communication error has occurred. The glitching of games like *Mario Kart 8* is on one level relatively inconsequential, yet on another level reveals the complex infrastructural entanglements of what might be called the geopolitics of latency.

The geographical distribution of server regions is a political issue. According to the developer Subspace 'three quarters of all internet connections in the Middle East are outside playable latency levels for dynamic multiplayer games,' compared to a quarter in the United States and Europe (O'Halloran, 2022). This is not simply a matter of server location nor an issue of broadband connection but of a global disparity of speed. For example, it is predicted that in Latin America, by 2026, '44% of the population will have access to fixed broadband services... but only 5.3% will be on a connection delivering 500Mbps or more, and only 1% will have speeds of more than 1Gbps' (O'Halloran, 2022). As described above, these kinds of hardware solutions (exosomatic) that attempt to overcome the volatility of latency are geopolitical in that they are physically located in a geographical region. However, infrasomatically they must also be thought in relation to the body (endosomatic). For example, the location of a data centre or the specification of a fibre optic cable can mean that data can be extracted from the body and reach its destination quicker.

To think the volatility of latency beyond the endo and exosomatic as infrasomatic is to understand it not simply as a geopolitics of the earth—the location of a data centre or the specification of a cable or server. It is to think latency as a geopolitics of the body, which is to consider the body as infrastructure. How quickly can the valuable data contained in the single firing of a neuron travel to its destination? How quickly can that information be processed by a machine learning algorithm? At what speed can the results of that processing be redeployed? How quickly can wants, needs, desires, and impulses be produced?

These are questions of the infrasomatic nature of latency as a geopolitics of the body. An example of this is a software solution called Rollback Netcode deployed in some online one-on-one fighting games in the form of a predictive algorithm called Good Game Peace Out (GGPO).

GGPO

GGPO uses delay-based netcode (for example, a three-frame delay applied to both players). To account for latency spikes that exceed the delay; however, Rollback allows the game to continue by predicting what the next move might be. For example, a player in Dublin inputs a command to punch their opponent, this appears on their opponent's screen three frames later and the game runs smoothly. However, what if there is a spike in latency and the players are now out of sync? The player in Dublin does not receive their opponent's next move. Rather than the game glitching or stopping, GGPO predicts what the player in Boston's move is going to be. The game continues with the predicted move. Five frames later the Boston based player's move arrives in Dublin. If the prediction was correct the game continues. If the prediction was not correct the game rolls back five frames and replays the move based on the actual input. In short, GGPO simulates gameplay, predicting the future, to adjust the past to change the present while maintaining a coherent experience of the present. This is an entirely different kind of entanglement between the technological object and the body when compared to the example of the blind person's cane. The circuit of subjectivity gets short-circuited by an algorithm that operates below the threshold of human perception yet produces a particular way of seeing the world.

GGPO works by converting objects in the computer's memory into a format that can save different game states and load them. This process is called serialization. For Rollback Netcode to work successfully two things must be in place: Serialization, and the separation of game logic from game rendering. Firstly, to be able to roll back and correct the game, every single frame of gameplay must be serialized. It must be converted into a storable stream of bytes that can be recalled and rendered when needed. Secondly, the game logic—simply put, the code that runs the game—must be separated from the rendering of individual game frames. This means that in theory the game could continue to run without any of the gameplay being rendered and could do so at speeds below the threshold of human perception, where different game states and game logic are being continually simulated in the background (Pusch, 2019). Importantly, when the game rolls back to a frame where a prediction error was made, what is seen on the screen is not a rendering of the actual move but a resimulation of a move that was previously made. Importantly therefore, Rollback is not rendering, but resimulation of an actual past that was serialized but not rendered, and is resimulated in the future in order to correct the past.

This resimulation produces what Shane Denson calls 'discorrelated images.' These are images that operate below the threshold of human cognition. They call into question the relationship of the subject and the object, and therefore ideas about how we perceive the world (Denson, 2022a). This is important as we can only know the world through images. These may be literal images like photographs, paintings, or Instagram posts. They may also be images in the sense of impressions that we have about the world. Impressions that may be produced by our social context, politics, gender, status, etc.

In our age of contemporary algorithmic technology, these representations of the world are becoming increasingly automated. If technological representations of the world are the condition of our experience of everyday life, then changes in these systems of representation have the possibility to significantly change our experience of everyday life.

For example, during the Black Lives Matter protests of 2020, techniques of algorithmic governance were used to produce certain kinds of space. Multiple US police departments employed the services of the company Dataminr who offer 'real-time AI for event and risk detection… [by detecting] the earliest signals of high-impact events and emerging risks from within publicly available data' (Dataminr, n.d.). By publicly available data, what they actually mean is that X (formerly Twitter) has given them access to their Compliance Firehose API which scans the archive of every public tweet as soon as its author hits send. That information can then be accessed using Dataminr's interface which, on one side of the screen, lists Tweets and on the other the location of those tweets—as well as numerous other pieces of information. Algorithmic tools of automated image representation like these are increasingly used to produce certain spaces of governance in the city.

Between the act of producing a representation of space, in the form of image code, and the representation of space itself as code image, a space is opened out for image manipulation to take place at speeds that often operate below the threshold of human consciousness. In GGPO, this so-called discorrelation occurs mostly below the threshold of human perception. However, on the occasion where a significant spike in latency occurs and the game must roll back a number of frames some glitching can occur. A player who was predicted to be walking forward but had in fact stopped walking will suddenly jolt back and the gameplay will not appear smooth. In terms of gameplay this glitching is usually inconsequential. This glitching shows how Stiegler's foundational concept of the co-constitutive nature of the human and technology is becoming increasingly more complex in our age of contemporary algorithmic technology. It highlights how technologies are mediating our perception of everyday life by producing technologically manufactured temporalities. It dispels the myth of the immateriality of the network by taking us beyond the sleek black screens of our luxury devices, and highlights the entangled web of physical infrastructures and digital codes that facilitate everyday life.

A Question of Hope

This chapter has introduced a series of key interconnected concepts in the work of Bernard Stiegler—the co-constitutive nature of the human and technology and the endo and exosomatic. The co-constitutive nature of the human and technology is foundational to all of Stiegler's work. It describes how the evolution of the human cannot be thought apart from technology. The concept of organology, briefly introduced at the beginning of the chapter, situates the technological evolution of the human across the multiple levels of society. The concept of pharmakon, described more fully in Chapter 5, shows how technology has either a toxic or curative effect across these levels and asks the question of the extent to which it might be possible to find a cure or a therapeutics. Using the example of the algorithm Good Game Peace Out, I showed how the relationship between the endo and exosomatic evolution of the human is complicated by technologies that operate below the threshold of human consciousness.

The question of the extent to which hope is possible in our age of algorithmic technology is a question of the extent to which it is possible to understand both the positive and negative effects of technology.

Using the concept of pharmakon, Stiegler describes these effects as either the poison or the cure of technology. He calls for tactics to be prescribed that could be a kind of therapeutics, capable of acting across the multiple scales upon which these technologies operate. To understand the extent to which a therapeutics is possible in the infrasomatic city is to understand changes to the exosomatic exteriorization that intervenes in human experience prior to the 100-millisecond range where perception and sensation register (Denson, 2022a). It is not just to understand how these technologies mediate our experience of the world, but how biological and technical organs are becoming increasingly intertwined. To find a therapeutics is also to ask to what extent the human body is becoming a component in the network of algorithmic infrastructures. What evolutions such as new neurological flows and ecologies of attention and addiction are being produced (Citton, 2017)? Stiegler investigates this question by exploring the effect that digital technologies have on memory. This is the subject of the next chapter, which explores Stiegler's ideas about the technological production of memory by introducing his concept of tertiary retention.

CHAPTER 3

Stiegler and the City

Introduction

The previous chapter introduced Stiegler's concepts of 'exosomatization' and 'endosomatization' and the 'co-originary' status of technology and human evolution (Berry, 2016). As a reminder the endosomatic is that which belongs to an individual organism from birth—a limb, wing, eye, or finger—whereas the exosomatic includes technologies that are not part of that individual from birth. This special relationship that the human has with exosomatic organs, like the blind person's cane, means that the human is essentially a technical being and to be human means to 'exceed the biological' (50). The human is essentially defined by '[t]he evolution of the "prosthesis"' and that prosthesis is technical (Berry, 2016). In Chapter 2 I proposed that in our contemporary technological age the endo and exosomatic have become further complicated by the entanglement of technological infrastructures and the body through algorithmic technologies. I contend that this is a matter of the extent to which the right to the city is curtailed and should therefore be a matter of consideration for architects. I argued that the complex multi-scalar infrastructures of our contemporary technological condition create evermore entangled circuits of subjectivity between human and non-human cognitive agents (Bigo et al., 2019, p. 48). This chapter proposes that the concepts of endosomatization, exosomatization, and infrasomatization are useful for thinking the city in our age of contemporary algorithmic technology.

Stiegler and the City

Stiegler and the Internation Collective (2021) use the concepts of endo and exosomatic as a way of thinking the city. This relates to philosopher and sociologist

Henri Lefebvre's theory of the city, understood as a complex interconnected process of urbanization. Lefebvre contrasts the idea of the city as a bounded geographic location with an idea of the urban as an interrelated process, as an 'ecosystem,' or a 'social reality made up of relations' (Lefebvre, 2008, p. 72, 103). For Lefebvre, the process of urbanization is the aftershock of the implosion/explosion of the city. The city is no longer contained within the boundaries of a wall as was the case in a typical medieval city. Rather, as trading, communication and transportation developed, the city began to 'explode' beyond the city wall. Moving beyond city walls, the distinction between rural and urban was transformed, like a 'fabric thrown over a territory' (71). As capitalism developed, the process of urbanization no longer respected borders and began to subsume territories. The city still exists. It is not the opposite of the urban, rather it has become caught in the net of the fabric of urbanization and 'is no longer lived, ... no longer understood practically. It is only an object of cultural consumption for tourists, for aestheticism, avid for spectacles and the picturesque' (Lefebvre, 2008, p. 148).

Lefebvre detects a shift in what a city is for. The city is still a place to live, but the city was also becoming an object or a product among other products in a capitalist economic system that could be traded and branded and represented in particular ways for the benefit of the few and not the many. The city entered into a system of exchange. The city's use value was now becoming dominated by its exchange value. To put it another way, it was becoming no longer a place for the many to dwell but a product for the few to trade. To describe the development of the city, Lefebrve imagines a scale of urbanization. Along this scale of urbanization, he places three different types of city—the political, commercial and industrial town, followed by a final category that he calls the 'critical point' (123). In *The Urban Revolution* (2003) this scale moves from the 'political city' to the 'mercantile city,' transitioning from the 'agrarian to the urban' into the industrial city reaching the 'critical zone.' This is the process of the city 'exploding' (Lefebvre, 2003, p. 15).

It could be argued that, in our age of contemporary digital technology, a new critical point of urbanization has been reached.

This so-called 'smart' urbanization is underpinned by what geographers Rob Kitchen and Martin Dodge call 'Codespace.' Codespace is where '[s]oftware is... bound up in, and contributes to, complex discursive and material practices, relating to both living and non-living, which work across geographic scales and times to produce complex spatialities' (Kitchen and Dodge, 2011, p. 13). In a similar way to the concept of infrasomatization that I introduced in Chapter 2, Codespace is Kitchen and Dodge's way of trying to explain how contemporary technology creates new kinds of spaces. These new complex spatialities are produced by what architectural theorist Keller Easterling calls 'extrastatecraft.' She refers to the process by which '[s]ome of the most radical changes to the globalizing world are being written, not in the language of law and diplomacy, but in... spatial, infrastructural technologies' (Easterling, 2016, p. 16). How are we to think these changes in relation to the city? The following section applies Stiegler's concept of exosomatization to the city. I then introduce the concept of algorithmic governance, and propose a way of thinking about the city based on an advancement of Stiegler's concepts of endo and exosomatization. I call this next stage in the process of urbanization 'the infrasomatic city.'

Exorganic City

I suggest that to understand the significance of algorithmic governance and the city in our age of contemporary digital technology is to understand the body as part of the network of technological infrastructure. I argue that to do so is useful for understanding urbanization in our age of contemporary digital technology.

It is useful because it foregrounds that the city, which Stiegler describes as an 'exorganic milieu,' is composed of infrastructures that operate simultaneously across multiple scales from the micro-biological to the planetary. The city, according to Stiegler, is made up of 'aggregates and localizations of complex exorganisms within which simple exorganisms cooperate' (Stiegler and the Internation Collective, 2021, p. 67). Stiegler applies the term exorganic to the city because the city must be understood through the technological infrastructures that produce it. By milieu, Stiegler, following philosopher Gilbert Simondon, means the external environment that produces individuals where those individuals produce the external environment. By exorganic Stiegler means that the human inhabitants of the city are themselves technological infrastructures. Stielger understands these technological infrastructures as not simply appendages added onto humans but as being constitutive of human individuals (Stiegler and the Internation Collective, 2021, p. 67). With Stiegler, I argue that the city must be understood as a network of biological and artificial organs that cannot be thought as separate from its inhabitants but is in fact constitutive of them. The idea of the city as exorganic is further complicated by our age of 'algorithmic governmentality' (Rouvroy, 2020; Rouvroy and Berns, 2013; Rouvroy and Stiegler, 2016).

Algorithmic Governmentality

The idea of the city as a complex exorganism is not specific to our age of contemporary digital technologies. Artificial organs such as roads, rail networks, sewage systems etc., are also exosomatic organs (Berry, 2016). However, in our age of digital contemporary technology, these exosomatic entanglements have become increasingly more complex as algorithmic technologies are deployed in urban governance. Antoinette Rouvroy (sometimes with Thomas Berns) calls this algorithmic governmentality. Algorithmic governmentality describes the governance of 'the social world that is based on the algorithmic processing of big data sets rather than on politics, law, and social norms' (Rouvroy, 2020). Rouvroy follows philosopher Michel Foucault's understanding of governmentality as 'the ensemble formed by institutions, procedures, analyses and reflections, calculations, and tactics that allow the exercise of... power that has the population as its target, political economy as its major form of knowledge, and apparatuses of security as its essential technological instrument' (Foucault, 2007, p. 108).

Put simply, Foucault uses this concept of governmentality to understand how bodies are produced as docile by a range of different forms of governance beyond simply political rule. However, with her concept of algorithmic governmentality, Rouvroy wants to turn Foucault on his head and show that, in a digital age 'it is no longer a matter of producing docile bodies according to a norm—medical, pedagogical, carceral etc.—but of making the norms docile according to the body' (Rouvroy and Stiegler, 2016, p. 23). By producing docile bodies, Foucault shows how methods of government and governance make us ready to accept rules and laws. Rouvroy argues that the body now plays a very different role in producing such norms.

This is a complex idea but perhaps a simple way to explain it is the now familiar phrase and its various permutations from online shopping websites: 'You bought this so we thought you'd like this.' The consumer desires of the body produce a norm which in this case is a consumer experience tailored to produce further desires, wants, and needs. In this sense, algorithmic governmentality is infrasomatic. It foregrounds the entanglement between technological infrastructures and the body. Rouvroy outlines three stages of algorithmic governance.

The first of the three stages are the collection and automated storage of unfiltered mass data. Second, the automated processing of these big data identifies subtle correlations between them, which is a matter of knowledge production. Third is the deployment of data to anticipate individual behaviours and associate them with profiles defined on the basis of correlations discovered through data mining (Rouvroy and Berns, 2013). Algorithms, along with roads, rail networks and sewage systems can thus be imagined as exosomatic organs; infrastructures that are integral to the functioning of everyday life in the city. However, as David Berry (2018) proposes, algorithmic infrastructures differ from non-computational infrastructures like roads. He argues that they are not simply exosomatizations in the Stieglerian sense that I have outlined above. Algorithmic infrastructures, suggests Berry, are 'not just the production of tools or instruments' but are 'the production of constitutive infrastructures' that combine and produce 'endosomatic capacities and exosomatic technics' (2018). Which is to say that they are not simply exosomatizations as the production and use of tools, but rather fuse together 'endosomatic capacities and exosomatic technics' (Berry, 2018). This means that algorithmic infrastructures are not simply tools for acting in or upon the world, or indeed tools in the Stieglerian sense of technical prostheses, but are themselves infrastructures that produce the very conditions for the possibility of thought and action. Berry calls this process 'infrasomatization' (Berry 2018). In his last published text, Stiegler seems to embrace this advancement of his concepts of endo and exosomatization (Stiegler and Internation Collective, 2021).

Infrasomatization

In an attempt to go beyond what he sees as the binary of endosomatic and exosomatic Berry proposes the term 'infrasomatization' (Berry, 2018). Drawing on the Latin infra, meaning 'below' or 'under a part of the body,' Berry defines infrasomatization as 'the capacity for framing or creating the conditions of possibility for a particular knowledge milieu' (Berry, 2016).

Put simply, there are infrastructures and networks that operate outside of human perception. These infrastructures and networks are increasingly governing everyday life and produce new forms of knowledge, ways of seeing the world, and ways of being in the world. It is in the infrasomatic city that the new complex infrastructural configurations of algorithmic technologies are creating new cognitive entanglements between technological infrastructures and the body.

Therefore our current urban condition is marked by 'social structuring technologies that inscribe new forms of the social (or sometimes the anti-social) onto the bodies and minds of humans' and create 'cognitive infrastructures that proletarianize our cognitive faculties' (Berry 2018).

Such infrasomatic infrastructures, suggests Berry, are 'not just the production of tools or instruments' but are 'the production of constitutive infrastructures' that combine and produce 'endosomatic capacities and exosomatic technics' (2018). This means that algorithmic infrastructures are not simply tools for acting in or upon the world, or indeed tools in the Stieglerian sense of technical prostheses, but are themselves infrastructures that produce the very conditions for the possibility of thought and action. As such, algorithmic governance is, suggests Rouvroy, a matter of the 'colonization of public space by a hypertrophied private sphere' (Rouvroy, 2020; Rouvroy and Berns, 2023). It colonizes public space by 'producing' peculiar subjectivities. Fragmented, 'the subject comes in the form of a myriad of data that link him or her to a multitude of profiles (as a consumer, a potential fraudster, a more or less trustable and productive employee and so on)' (Rouvroy and Stiegler, 2016). Therefore if algorithmic governance is a matter of the production of public and private space, it is also a matter of the extent to which a right to the city is possible in the infrasomatic city.

For example, the algorithmic governance of everyday life is being employed by Google/Alphabet—as well as others like Cisco and IBM—who, in partnership with city authorities, are seeking to deploy various data gathering technologies in the urban realm. These include their now cancelled Sidewalk Labs project, an urban regeneration scheme that would include the deployment of machine learning and data gathering technologies (Capener, 2018). A word that has become synonymous with these kinds of projects is 'smart.'

The term 'smart city' is problematic in that it glosses over the complex social, cultural, ethical and political issues that many of the technologies deployed raise. It also suggests that the city and its inhabitants are somehow smarter because of the deployment of so-called smart technologies.

Stiegler counters the idea of the smart city by arguing that these kinds of technologies also produce a kind of stupidity, in that their automation removes the need for thinking. This is another example of what Stiegler calls the pharmacological nature of technology—that technology has the potential to be both toxic and curative.

A Smart City?

As early as the 1990s, the term smart city was used to refer to urban centres that were key political, economic, and cultural hubs or 'technologized coordinating centres for the flow of goods, services, communications, and people' (Halegoua, 2019, p. 28). In the 2000s, IBM began using the term 'smart cities' to describe 'how various public services and infrastructure projects can be enhanced with information technology and data analysis' (Halegoua, 2019, p. 26; also see Back, 2009). Today, IBM along with companies like Siemens, Cisco and Google/Alphabet are all engaged in the design and implementation of so-called smart city technologies. In 2014 IBM designed and provided technological infrastructure to the city of Rio de Janeiro, including a large control centre that integrates data from 30 different agencies for use by the city's police. This system has been criticized by Amnesty International for the number of people killed since its deployment: 182 people, a 78% increase (Capener, 2018).

A significant recent development in the smart city discourse was the founding of Sidewalk Labs, Google's urban innovation unit. Sidewalk Labs is significant because the company's purpose, as articulated by founder and CEO Dan Doctoroff in 2016, was stated as the replication of 'the digital experience in physical space.' They proposed to use various data gathering AI, machine learning, and sensing technologies 'including cameras and location data as well as other kinds of specialized sensors.' The deployment of these technologies would be funded 'through a very novel advertising model… We can actually then target ads to people in proximity, and then obviously over time track them through things like beacons and location services as well as their browsing activity' (Zuboff, 2019, p. 230–231). What Doctoroff is describing is not unique to Sidewalk Labs' project. Facilitated by personal mobile technologies, this is now a significant present and future business model of everyday life in our contemporary digital age. The infrastructures that facilitate it often operate outside of human perception infrasomatically below the surface of everyday life.

Another example of how algorithmic technologies mediate and influence everyday life is the game *Pokémon GO*. Using gamification coupled with automated image production, *Pokémon GO* influences how users travel around a city, where they go and where they might spend money. The forerunner to *Pokémon GO* was a similar location-based game developed by Niantic Labs and Google, called *Ingress*, which was itself based on an earlier Niantic game called *Field Trip*. The example of Ingress is useful because it foregrounds many of the processes present in the algorithmic production of space in everyday life. Ingress connects with Google Maps and Streetview as well as having a video series on YouTube which emails your Gmail account with updates, is largely organised on Google Hangouts and G+, as well as collecting data to further enhance Google Maps' ability to provide the best walking routes (The Guardian, 2022).

These processes operate across multiple organological scales— the biological, technical, and social as well as geographic scales—meaning both the earth, but also in the etymological sense of earth writing; the production of territories. Because of the access that these technologies have, and the speed at which they operate, these kinds of technologies are able to move simultaneously across the organs of society in ways that were previously not possible.

It could be argued that that the digitally augmented and/or virtual mediation of reality will play an increasingly important role in the infrasomatic city. Companies like Niantic Labs have made their *Lightship Augmented Reality Developer Kit* (ARDK) open source, allowing developers to use the same technology that produced *Pokémon GO* to develop other space-based augmented reality applications. The company that claims to be 'inventing the augmented world of tomorrow' is also developing real-time machine learning, sensing, seeing, and mapping technologies which they hope will be able to produce and mediate the real-time representation of everyday life (Niantic Labs, 2022).

The concept of infrasomatization, developed by David Berry, brings together Stiegler's concepts of endosomatization and exosomatization to understand how technological infrastructures are creating new forms of governance by producing new closed loops between the body and algorithmic infrastructures (Berry, 2016). The concept of infrasomatization is productive because it foregrounds the idea that algorithms are constitutive infrastructures. They are structured by, and structure, the body. They produce new circuits of subjectivity between the body, everyday life, and platform infrastructures. Stiegler calls this the 'becoming-mnemotechnical of every material, substance and product' (Bigo et al., 2019). By mnemotechnical, Stiegler is referring to the way that technology does not just aid memory but is constitutive of it.

It is important to highlight the collective nature of infrasomatic processes. Stiegler proposes that the infrasomatic process of the becoming-mnemotechnical of every material, substance, and product takes place collectively.

<u>Algorithmic technologies can be understood as 'social structuring technologies that inscribe new forms of the social onto the bodies and minds of humans' (Berry, 2018).</u>

Stiegler proposes that human memory is not something that is produced individually but is produced collectively. He also proposes that human memory is not just a biological function of the human body (endosomatic) but is also produced by technological prostheses (exosomatically). Stiegler argues that devices like smartphones are constitutive of memory and as such are cognitive infrastructures. He argues that they do something to our cognitive faculties that has a negative effect, which is that they proletarianize us. By proletarianize, he means that—through our use of contemporary digital technologies—while we accumulate knowledge, we also lose access to the knowledge. This is one of the pharmacological aspects of contemporary digital technology. We accumulate certain kinds of knowledge like watching a YouTube video on how to fix a leaking pipe. However, we also lose knowledge, which for Stiegler means losing the knowledge to create, collaborate, and self-organize. By drawing together the endo and exosomatic into an increasingly closing circuit that passes between the body, the brain, and various technological platforms, the concept of infrasomatization foregrounds how processes of proletarianization increasingly operate below the threshold of human consciousness. For example, it could be argued that machine learning algorithms, like the type used in ChatGPT and other AI text generation software that can produce large amounts of text at speeds faster than human cognition, create a loss of knowledge and creativity because they automate much of the process of thinking. David Berry, following Stiegler, calls this 'hyper-proletarianization'—the weakening and potential annihilation of human reason creating 'anti-thought,' a kind of collective stupidity (Stiegler and the Internation Collective, 2021, p. 78).

To think algorithmic infrastructures infrasomatically is to understand them as not simply tools in an exosomatic sense—although they are part of exosomatic prostheses like a smartphone etc.—but as constitutive infrastructures that produce the very conditions for thought and action. They have potential to produce modes of automated cognition through algorithmic governance. These processes of automated cognition 'become hegemonic through a colonization of an increasing number of social spheres, from law, politics, communication, media and education, etc.' (Berry, 2018). I argue that it is useful to think of the city as infrasomatic. This creates a framework from which the complexities of the city in our contemporary technological condition might be thought.

To think the city infrasomatically is to describe the way in which our age of algorithmic governance increasingly closes the circuits between the body and digital technological infrastructures, producing multi-scalar circuits between the body and the infrastructures of everyday life. To begin to understand the extent to which a right to this kind of city might be possible in our contemporary technological condition is to understand the city as infrasomatic—as a network of artificial organs that cannot be thought as separate from its inhabitants but are co-constitutive with them.

CHAPTER 4

Technology and Memory

Introduction

Chapter 2 introduced a foundational concept in Stiegler's philosophy, the co-constitutive nature of the human which is to say that the human does not just invent technology, but humans are an invention of technology. The concepts of endo and exosomatization were introduced using the example of a blind person's cane. A key idea was introduced: Technology does not exist apart from the human body but is part of a circuit of the production of subjectivity—our self-awareness and attitudes to the world around us. It was argued that algorithmic technologies have complicated this relationship because they have access to biological organs in ways that technologies like the blind person's cane do not. To describe this relationship, Chapter 3 introduced the concept of infrasomatization which describes how algorithmic technologies produce new networks of cognitive agents which commodify human capacity for reason and thought (Bigo et al., 2019). Using the example of the algorithm Good Game Peace Out, it was shown that contemporary digital technologies have the potential to mediate everyday life in ways that operate 'prior to the 100-millisecond range where perception and sensation register for humans' (2019). Furthermore, these technologies do not just mediate our experience of the world, they are entangling endosomatic biological organs together with exosomatic prostheses in new ways. The body thus becomes an integral component of algorithmic infrastructures producing endosomatic evolutions such as new neurological flows which themselves produce new ecologies of attention and addiction (Citton, 2017).

Further developing the concepts presented in Chapters 2 and 3, this chapter introduces a question central to Stiegler's work, the extent to which endosomatic technologies mediate and produce our perception of everyday life. One of the ways that Stiegler tries to think through this question is by considering the effect that contemporary digital technologies have on memory. He argues that our technological prostheses do not just aid memory but that they *are* memory (Stiegler, 2009b, p. 60). Stiegler calls this tertiary retention. This term refers to memory that exists outside of the body in material form, yet constitutes memory that is internal to the human body. Because memory here exists outside of the human body, this also means that tertiary retention is spatial. Human memory is not passed down from generation to generation genetically. Instead it is passed down outside of the human body, through external material like a sharpened flint, a stone tablet, a book, a text message, or a tweet. Importantly for Stiegler 'tertiary retention is… spatial' (Stiegler, 2011, p. 73). The co-originary status of the human with technology, and the production of memory via technical artefacts, means that the production of memory is spatial. Because, for Stiegler, technology and the human are co-constitutive, this also means that forms of retention external to the human body are also co-constitutive of forms of memory that are internal to the human body.

This chapter introduces Stiegler's concept of tertiary retention by using the example of the gramophone. This illustrates Stiegler's basic framework for understanding how retention works. It is, however, limited when trying to understand how contemporary algorithmic technologies complicate the processes of retention. A useful example for illustrating the way that contemporary digital technology creates complex circuits of retention is the iPhone camera. The camera acts as a form of tertiary retention because it is used to compile a memory repository of images that exist outside of the human body and nonetheless form part of human memory, but in doing so fundamentally changes the nature of what an image is and does. Before introducing the concept of tertiary retention in more detail I set the context for Siegler's philosophy of retention by introducing Anthropologist André Leroi-Gourhan whose work Stiegler draws on, in particular by returning to the ideas of the co-constitutive nature of the human and technology and how memory is produced.

Memory Programs

One of the primary influences on Stiegler's philosophy is Leroi-Gourhan's book *Gesture and Speech* (1993). Like Stiegler, Leroi-Gourhan argues for the co-constitutive nature of the human and technology, and how memory is exteriorized in material objects. For Leroi-Gourhan this process of exteriorization happens through rhythms produced by certain gestures. These objects become the media of recording for the rhythms that produce them—a kind of media for storing certain kinds of memory (During, 2017, p. 50). For example, a prehistoric sharpened flint, while not made to retain memory, becomes a 'vector of memory' (50). By vector of memory, Leroi-Gourhan means that the object has the potential to carry memory from generation to generation. It is for this reason that archaeologists are able to reconstitute a civilization through the study of objects not simply through their use value as a record of human motor function, but also as a recording of human behaviour and human spirit (59). Other forms of media like cave art, Amerindian cords made of knots and tattoos, and early numerical and alphabetical systems are also forms of the materialization of human bodily rhythms, be they neurological flows or repetitive muscular gestures. All these are examples of the materialization of memory.

Gourhan talks about rhythms which he divides into three categories. Firstly, rhythms are expressed biologically in the body. For example, the rhythm of a heart or breathing. Secondly, rhythms are exteriorized in bodily gestures such as singing and dancing which are essential to the formation and cohesion of cultural groups. Thirdly, rhythms are exteriorized in tools, which is itself a process that is constitutive of the human. Gourhan develops a nuanced and interconnected 'rhythmology' by connecting the production of rhythm with the activation of memory through technics—'[t]echnics does not aid memory, it is memory' (Stielger, 2009, p. 65). Stiegler calls the study of these memory programs 'programmatology' (65–96). By program, Gouhran and Stiegler are referring to operational sequences that are used in different ways according to the kind of memory activated (Audouze, 2002, p. 277–306). For example, the memory activated by a tool such as a flint produces a specific series of operations like cutting or carving. These tools require certain bodily gestures that shape the body in certain ways as muscles are produced, as well as new kinds of knowledge about the extent to which the tool can be used and the materials it can be used on.

Corresponding to each memory program is what Stiegler calls a 'rhythmics.' These are imagined to be exteriorized in programs that are 'no longer inscribed in the organism itself' (Stiegler, 2009b, p. 70–71). However, by thinking Stiegler's rhythmics infrasomatically and spatially it could be argued that the nature of our contemporary technological condition means that, while memory programs are exteriorized, they are also reinscribed back into the organism itself in the form of shaping the body physically like the example of the flint above. Gouhran distinguishes three different but interconnected kinds of memory: Specific or genetic, individual memory, and socio-ethnic memory (70–71).

Firstly, specific or genetic memory can be understood as memory shaped by individual species-specific experiences such as external stimuli; its rhythmics is 'physiological.' Which is to say that memory is related to biological nature and is activated in operational sequences that involve acts such as feeding and sex. Secondly, individual memory, present in more complex mammals, is produced through experience and education. It is accumulated through knowledge and transmitted and preserved by language. Its rhythmics is 'figurative.' This means that in its accumulation it becomes semiautomatic and habitual, not requiring one's full attention like driving a car or brushing your teeth. Thirdly, socio-ethnic memory is the collective memory of a particular cultural group. Its rhythmics is characterized as 'functional,' along with the knowledge transfer of individual memory. Its 'repetition guarantees the subject's normalized equilibrium within the social milieu, or social setting' (73).

<u>For example, the repetition of certain rhythms is reliant on the stability of the calendrical structure of a given epoch. It is the repetition of such daily, monthly, annual, cycles that produce the</u>

rhythms of everyday life—work, eating, hospitality etc. These rhythms are an 'indispensable aspect of group cohesion, as essential for the group as for the individual; its disappearance would mean the destruction of ethnic unity' (Stiegler, 2009b, p. 70–71).

It is important to note that by ethnic unity, I understand Stiegler to mean cultural unity, not ethnic unity in the narrow sense of a specific ethnicity. These rhythms, Stiegler argues, cannot be thought separately from the technologies that produce them. These three programmatic levels of memory mean that processes of individuation within a given group, in a given epoch, are fragile. The possibility is always present for technics to change culture (65–96). This study of the production of memory programs is what Stiegler, following Leroi-Gouhran, calls 'programmatology' (Stiegler, 2010b, p. 35; 2009, p. 69).

Grammartization

It is important to note that Stiegler does not consider these forms of exterior memory as simply totems, used as just memory aids, but they are constitutive of memory. This is a process that Stiegler, following philosopher Jacques Derrida and linguist Sylvain Auroux, calls grammartization. This is the exteriorization of memory in the form of marks and traces. It is important to note that the history of the city is a history of new forms of grammartization. For example, the invention of the alphabet creates the possibility of the Greek city-state. Without the possibility to record, document and distribute the laws of urban life, the city could not function (During, 2017, p. 69). In the same way, much of modern everyday life in the city could not function without forms of grammartization. For example, the codes and algorithms that facilitate camera phone technology. For Stiegler, these kinds of endosomatic technologies produce new ways of being in the world because they act as forms of tertiary retention. They produce a kind of memory that exists externally to the human body, but constitute biological memory. For Stiegler this kind of external memory cannot be separated from biological memory. They are not two separate and distinct types of memory—they are memory. The following section introduces the concept of tertiary retention in more detail.

Tertiary Retention

Tertiary retention is memory storage that is external to, but constitutive of human memory. Familiar examples would be the iCal app on the iPhone, a Facebook timeline, or an X (formerly Twitter) feed. This third kind of memory, exterior to the human body, and therefore spatial, aids the recall of specific memories. I no longer need to remember the meeting I have tomorrow because my calendar will remind me. Using the example of a turntable, Stiegler writes: 'You only have to listen twice to the same melody to see that between the two auditions, consciousness (the ear, here) never hears the same thing... because the ear of the second audition has been affected by the first' (Stiegler, 2011, p. 3).

That I can perceive the same melody twice (or look repeatedly at the same image on Instagram), but each time experience something different, means that my perception of the melody (or the image) is not constituted by primary retention, but is in fact constituted by what Stiegler calls tertiary retention (Roberts, 2006, p. 55–63). Tertiary retention devices are spatial and digital. This is of great significance for spatial production in our age of contemporary digital technology. It means that the possibility exists for their manipulation (Stiegler, 2011, p. 73). The speed at which digital tertiary retention devices operate, the access they have to our immediate experiences—where we go, what we buy, what we think, what we like—and their ability to record, remember, and process events in real time changes everyday life.

Tertiary retention is the storage of memory in memetic devices that are external to the human body. Stiegler bases his ideas about retention on the work of philosopher Edmund Husserl. For Husserl, retention comprises two parts: Primary and secondary retention. Using the example of listening to a melody, primary retention can be understood as the perception of what is happening immediately around us. To make sense of a melody I must be able to retain and remember the note that I am immediately hearing as well as the note that precedes it. Secondary retention is the recalling of past memory. For example, the memory of a melody that I heard yesterday (Stiegler, 2009b, p. 37). However, something different happens when the melody is recorded, and I listen to it on a record player rather than hearing it performed live. This is what Stiegler calls tertiary retention.

Tertiary retention is a third kind of memory that is exterior to the human body—like a picture or a photograph that aids the recall of a specific memory. Husserl calls this kind of memory image consciousness.

However, Husserl's idea is very different from what Stiegler means by tertiary retention (Husserl, 2005). For Husserl, primary retention is different from secondary and tertiary retention. This is because primary retention is the means by which we directly perceive the world. Secondary retention is different because, rather than being an act of perception it involves acts of imagination. Secondary and tertiary retention thus come from primary retention. For Stiegler technology structures perception. Rather than primary retention constituting secondary retention, here it is tertiary retention that constitutes primary retention. Using the example of listening to a pre-recorded melody Stiegler writes:

> You only have to listen twice to the same melody to see that between the two auditions, consciousness (the ear, here) never hears the same thing: something has occurred. Each new audition affords a new phenomenon, richer if the music is good, less so if not, and that is why the music lover is an aficionado of repeated auditions—a variation of selections... From one audition to the next the ear is not the same, precisely because the ear of the second audition has been affected by the first. (Stiegler, 2011, p. 73)

This third memory is both supported and constituted by technological supports (During, 2017, p. 51). This is, literally, as old as time itself. Neither are possible apart from the technologies that produce them, beginning with the ability of prehistoric beings to sharpen flint, which for the first time in the history of life meant that the possibility now existed to 'transmit knowledges that were individually acquired, but by a way that is not biological':

> **Man is a cultural being to the extent that he is also essentially a technical being: it is because he is surrounded by this tertiary technical memory that he can accumulate the intergenerational experience that is often called culture—that is why it is absurd to oppose technics to culture. Technics is the condition of culture in as much as it permits transmission. (During, 2017, p. 59)**

To bridge the chasm between *tékhnē* (technology) and *epistēmē* (knowledge) Stiegler returns to two key concepts that were introduced in Chapter 2—'exosomatization' and 'endosomatization' (Stiegler, 2018b). The relationship between the endo and exosomatic is complicated by the relationship that the body has with contemporary algorithmic technologies. In the infrasomatic city (Chapter 3), what can be imagined as technological lattices have become woven into the fabric of everyday life to such an extent that the body and technological infrastructures have become increasingly entangled. The body must now be understood as infrastructure. An example of the entanglement of algorithmic technologies of tertiary retention and the body is a technology that has become embedded in everyday life to such an extent that it may not be possible to imagine life (or image life) without it—the smartphone camera.

Automated Images

The example of the smartphone camera, in particular the iPhone 11 and subsequent versions, foregrounds how our perception and experience of everyday life is increasingly becoming governed by algorithmic technologies that operate below the threshold of human perception. As discussed above this complicates the relationship between the endo and exosomatic, mediating our experience of everyday life by producing new supply chains of perception. The iPhone 13 and 13 Mini, iPhone 13 Pro and 13 Pro Max, iPad Mini (6th generation) and iPhone SE (3rd generation) all contain Apple's 64-bit 'neural processing unit', the A15 Bionic chip (Tomshardware, n.d). The A15 chip is central to what Apple call 'Deep Fusion' (91Mobiles, n.d.). According to Apple, the chip can process '15.8 trillion operations per second.' For example, on the iPhone 11 (using the A13 Bionic chip), any image is not taken as one image but is rather a burst of nine shots. The nine shots comprise 'four fast exposure photos, four secondary photos and a single long exposure photo (91Mobiles, n.d.). Using Deep Fusion, the phone combines the images to produce the 'best' image. The 'best' here being an aesthetic category determined by Apple. Importantly, this process of real-time editing starts before the shutter button is pressed (Tomshardware, n.d).

The first eight shots are taken before the shutter is pressed. Once the shutter is pressed the iPhone takes one single long exposure shot (somewhere below 1/2 or 1/6th of a second). In the time it takes between the opening of the camera application and the final long exposure shot, the A13 chip has analysed 24 million pixels selecting the 'best' individual pixels from each of the nine shots to produce a single image. The whole process takes about one second. The A13 chip uses its predictive algorithm to sort through each individual pixel to produce an image that has the 'best contrast, level of sharpness, fidelity, dynamic range, colour accuracy, white balance, brightness… higher dynamic range, incredibly high levels of intricate details, excellent low light imaging, low levels of noise and… accuracy of colours' (91Mobiles, n.d.).

The new Google 6 Pro smartphone uses similar edge AI technology utilizing Google's Tensor chip which can remove unwanted objects, blurring, and using what the company call 'Real Tone' which, they claim, can 'accurately photograph all skin tones' (Google, n.d.). This claim foregrounds the controversy surrounding earlier versions of the phone that were not able to accurately represent darker skin tones and, as such, is a reminder of the 'exclusionary biases encoded into technological systems' (Denson, 2022a; Amoore, 2024). Google/Alphabet also owns Coral, a company that offers 'a platform of hardware components, software tools, and pre-compiled models for building devices with local AI.' The company also offers code examples on its GitHub page (GitHub, n.d.). Codes include 'Pose estimation,' 'Image recognition with video,' 'Object tracking with video,' 'Person segmentation with video,' 'Semantic segmentation,' 'Keyphrase detector,' 'Basic Image classification,' and 'Basic object detection' (Coral, n.d.).

Philosopher Roland Barthes described a photograph taken with a traditional shutter camera as being able to capture the 'this was' of time (Barthes, 1993). Smartphone cameras change this. The image taken by an Apple iPhone 11 is less 'this was' and more 'these were—maybe.' The singular image is a collection of pixels edited out from nine discrete micro-temporal moments: A collage of moments. It is a compilation of the movement of time, the framing of duration, the production of a kind of algorithmically produced deep fake image. It is a representation of a past that did not happen or did not happen quite like that.

Just like, for Stiegler, the tertiary retentional object allows one to experience a past that was not lived. Put simply, listening to a recording of a musical performance allows us to experience the music even though we were not present at the actual performance. In the real-time production of an image the process of Deep Fusion allows one to experience a past that feels like it was lived but in fact was not (Stiegler, 2011, p. 120–125). The image produced by Deep Fusion is the past not lived in visual form.

This is a process, writes Stiegler, where '[p]hotons become pixels that are in turn reduced to zeroes and ones on which discrete calculations can be performed' (Stiegler and Derrida, 2002, p. 152). In the example of the iPhone, this process of discrete calculations becomes a process of the production and manipulation of secondary retentions that take place in the moment when primary retentions are being formed. While primary retentions are being mediated via camera screens, those retentions are manipulated in real time. As the media theorist Fredrich Kittler writes:

> [t]he ability of digital media to store, process, and communicate levels of the real inaccessible to human perception comes at the cost of humans no longer being able to determine whether that which is allegedly processed by media is not in fact produced by them. (Kittler, 2017, p. 2)

As everyday life becomes increasingly mediated via automated images, it becomes increasingly more difficult to tell if an image is real.

The Automatic Everyday

This chapter introduced Stiegler's concept of tertiary retention. Using the example of the iPhone camera, I showed how the processes of the production of memory, and thus our experience of everyday life, gets complicated by contemporary algorithmic technologies. I showed that while the digital exteriorization of memory is not new, our contemporary technological condition has pushed it to a new stage (Blom et al., 2016, p. 35). As philosopher Yuk Hui puts it, 'spatiotemporal distance between those recalling and what is recalled is collapsed, and a memory is iteratively reterritorialized in the moments of its recollection, over-determining it with the metadata of capture, storage and retrieval' (35). What Hui means by this is that, the quicker the processing of data becomes, the more possible it becomes for our perception of everyday life to be processed in real time. This is the automatic-everyday where we are sold the myth that our digital prostheses are opening out a world of possibilities when in fact, according to a specific and ever-changing grid of algorithmic governance, they may be closing it down. The automatic everyday is a process that simultaneously operates at multiple scales from the planetary to the urban to the neurological.

At an urban scale, the city has thus become a veil of secrecy that digital capitalism needs in order to survive. The currency of this new city is data (Halpern, 2014, p. 3). As the informational networks and feedback loops connecting us and our devices proliferate and deepen, we can no longer afford the illusion that 'consciousness alone steers our ships' (Hayles, 2017, p. 141).

Digital tertiary retention of this kind and scale is a global spatialization of memory that produces new modes of being in the world through 'a quasi-materialization of… consciousness,' altering our perception and conception of space. This produces new kinds of space (Stiegler, 2011, p. 73). The next chapter introduces a final concept from the work of Bernard Stiegler—the concept of pharmakon—which Stiegler uses to describe how technology has the potential to be both toxic and curative.

CHAPTER 5

The Poison and the Cure

Introduction

Readers will now be familiar with a fundamental question of Stiegler's work: To what extent do technics, particularly contemporary digital technologies, curtail the right to the city. Technology is the condition of our experience of everyday life, and our access to the world is always-already technically mediated. Therefore any change in the way the world is, what Stiegler calls a 'suspension of the world,' can only take place through changes in technological systems. So how might we understand technical systems, to the extent that we might be able to suspend worlds in favour of imagining and creating a life worth living (Stiegler, 2009a, p. 22). At the centre of these questions are two key concepts: Pharmakon and organology. This chapter investigates the concept of pharmakon in more detail.

Pharmakon and Organology

What Stiegler wants to explore through his concept of pharmacology is a question of how technology might be both toxic and curative, and how one might be able to find the therapeutic potential of a given technology. Coupled with this is the understanding that a therapeutics cannot be found in technology alone, but must be seen across the biological, psychic and social organs of society. As I introduced in Chapter 2, he calls this organology. For Stiegler, the question of pharmacology is a question of hope. It examines the extent to which, in our co-evolution with technical supplements, we might find new and better ways of being in the world. This is significant to architects because contemporary technological systems are increasingly entangling everyday life, and call into question the extent to which a right to the city is possible. Put differently, how can the practice of architecture become pharmacological in our age of contemporary digital technology. Imagining the practice of architecture as a pharmacology requires methodologies that are capable of understanding the multiple levels upon which contemporary algorithmic technologies operate.

Stiegler's Pharmacy

Stiegler uses the concept of pharmakon to describe how our technological condition has potential to be both a poison and a cure. What he means by this is that a technological prosthesis gives back what it takes away. Thus, it is both the remedy and the sickness. Code, in the form of an image recognition algorithm, can both detect cancer but also have a racial bias when the same technology is deployed in facial recognition. Stiegler takes the idea of pharmakon from the philosopher Plato. In his book *Phaedrus*, Plato uses the act of writing to show how—in Stieglerian terms—the technological prosthesis of a writing implement can be both poison and cure. The writing implement removes the need for memory practice and is therefore a poison. However, writing is also a cure because it is also a remedy to memory loss—in that a technological prosthesis (exosomatic) exteriorizes memory through the act of writing. Writing gives but it also takes away.

This is pharmakon. Just as a pharmacist dispenses a cure for an illness, the very same cure, should it be incorrectly prescribed or taken incorrectly, can become a poison. Inherent in the ingredients of the pill are both the poison and the cure. Stiegler argues that our age is a pharmacological age.

To adequately understand the technological processes at work is to think pharmacologically. Which is to say, it is to find and prescribe a therapeutics by which we might be able to counter the potential toxicity of technology.

Organology

The human is an invention. The human does not just invent technology, but the human is itself an invention of technology. As explored above 'the invention of the human is technics,' and 'technics is the condition of culture' (Stiegler, 1998, p. 137; see also During, 2017, p. 59). Without technology, there can be no human and cultural evolution. Technology must be thought as both endosomatic and exosomatic—that which is both internal and external to the human. Thus, when we think about the evolution of human bodies, we must not only think about biological organs but also technical organs. However, as I argued in Chapter 3, the term infrasomatic describes how the body becomes infrastructure as part of technological networks. The concept of pharmakon cannot be thought apart from the endosomatic, exosomatic, and now the infrasomatic and the multiple scales of the biological, social, and psychic organs of society across which the toxicity of our technological age works.

A Pharmacology of the Algorithm

To explore the concept of pharmakon in Stiegler's thought, the following section introduces a critique of the concept. The critique by media theorist Mark Hansen is useful for helping foreground the significance of the concept in Stiegler's work because it foregrounds how algorithmic technologies complicate the concept of pharmakon by producing new modes of experience that reconfigure the processes of retention. Hansen's critique suggests that Stiegler's concept of pharmakon is limited for understanding the toxicity of contemporary algorithmic technologies. I show that Hansen fails to see how Stiegler also considered the toxicity of contemporary algorithmic technologies that operate below the threshold of human consciousness. Hansen shows that the speed at which algorithmic technologies are now able to operate means that they are able to exploit what N. Kathryn Hayles calls the missing half second. This is the delay or latency between perception and consciousness. As the latency of computational networks becomes faster, more efficient, located closer to the body in the form of on-device machine learning—like the example of the iPhone camera given in Chapter 4—it becomes possible for sensations to be produced that are prior to consciousness, but also to produce protentions (imagining the future) in the form of desires that become essential to the functioning of affective capitalism (Hayles, 2017, p. 173). However, Hansen fails to recognize that Stiegler also understands that shifts in technological epochs change the possibility for the production of a therapeutics. I then suggest that Hansen could have addressed his criticism of *pharmakon* by thinking it alongside Stiegler's concept of 'general organology' (Stiegler, 2012, p. 12).

As global network latency (what might be called exo-latency) becomes more efficient, human biological latency (endo-latency) is affected because algorithmic technologies become able to change the order of perception. In exploiting the missing half second, these technologies can operate outside primary retentions and produce secondary retentions that may not have been lived. Thus, they have the potential to fundamentally change our experience of the present, our recollection of the past, and our possibility to imagine various futures. These light-speed technologies—and their real-time capture, storage, analysis, and re-presentation—instantly archive and process the everyday at such a speed that our experiences of everyday life no longer have time to take place (Ernst, 2017, p. 36).

The present is always-already archived, an 'algorithmicized present' (36). There is no longer a present-now of everyday life, only light-speed archival circuits that at the moment of archiving become, via algorithmic selection, possible futures (Amoore, 2020). If, as I argue, our only access to the world are representations of it and technological representations of the world are the condition of our experience of everyday life, then changes in these systems of representation—which I argue are becoming increasingly automated—have the possibility to

significantly curtail the possibility of the right to the city, which is to say a pharmacology of the city.

Building on Stiegler's work the question for Hansen is to what extent a pharmacology of such technologies is possible? Given the speed at which they operate, their complexity, and the multiple scales across which they simultaneously work, to what extent is it even possible to locate the toxicity and prescribe the cure that Stiegler seeks? Hansen recognizes a fundamental difference between a pharmacology of Plato's writing implement and contemporary algorithmic technologies. He argues that the pharmacological act of both remembering and forgetting is inherent in the act of writing and the technology of the pen—what the technology takes away it also gives. However, this is complicated when it comes to contemporary algorithmic technologies because as Hansen argues these kinds of technologies complicate the relationship between the poison and the cure so that they now '[require] a technical supplement *in order to make good on their pharmacological promise*' (Hansen, 2015, p. 50). For example, in order to make sense of a photograph taken on an iPhone, that iPhone requires a machine learning algorithm without which the image would not make sense. Sense making of the image is delegated to a machine learning algorithm.

This delegation is significant because the output of an image-based machine learning algorithm is directly related to the images upon which it has been trained. It is therefore by its very nature biased. Louise Amoore writes that '[a]t root, the algorithm can never be neutral or without bias or prejudice because it must have assumptions to extract from its environment, to adapt, and to learn. It is, ineradicably and perennially, a political being' (Amoore, 2020, p. 75). The results about 'who or what will be of interest, or who or what can be recognized' are already present in the assumptions, errors, bias, and weights. These results are fully political (69). If, as Hansen proposes, contemporary algorithmic technologies require a technical prosthesis to make good on their pharmacological promise—and, if these technologies are, as a requirement of their functioning, biased—then we are faced with a complex series of ethical issues (Hansen, 2015, p. 50). This is what happens between capturing a representation of an object and the representation of an object. Nicholas Mirzoeff (2011) calls this the 'the right to look'.

Thinking the Algorithm with Stiegler

That our technological epoch can be either toxic or curative is, for Stiegler, the defining characteristic of our age (Stiegler and the Internation Collective, 2021, p. 46). But how useful is the concept of *pharmakon* in *the infrasomatic city* where, as I have argued above, biological and technical organs are becoming entangled in more complex ways and are increasingly operating below the threshold of human consciousness? On this issue, Mark Hansen in his book *Feed Forward: On the Future of Twenty-First-Century Media* (2015) offers an interesting critique of Stiegler. Hansen's critique foregrounds the complexity of finding a therapeutics (a cure) but, I suggest, fails to understand the nuance of Stiegler's development of the concept in the transition from analogue to digital technological epochs. This transition is important because it marks the beginning of the increased entanglement between biological and technical organs—the endosomatic and exosomatic, what in Chapter 2 was introduced as the infrasomatic. The following section asks how pharmacology is possible in the infrasomatic city.

A Pharmakon in the Infrasomatic City

Hansen's critique is that Stiegler's *pharmakon* is unable to account for the complex ways in which contemporary digital technologies produce and mediate experience, at times below the threshold of consciousness. For him, the pharmacological structure of contemporary digital technology differs from writing and other analogue technical media. Writing, for Hansen, '*directly* gives back what it takes away, exchanging a "natural" source of memory for an "artificial" one' (Hansen, 2015, p. 50–51). In contrast, contemporary digital technologies 'involve an exchange of experiential modalities that also becomes an exchange of temporal scales of experience' (50–51). For example, Rollback Netcode (Chapter 2) exchanges the linearity of the experience of temporality with an algorithmically mediated perceptual regime that feeds a manipulated past forward into an adjusted present.

For Hansen, the mediation of experience via digital technologies is a new kind of pharmakon, both remedy and sickness. What is lost 'in the way of *perceptual* grasp of our environment,' he argues, we 'regain through an expanded and microtemporal sensory contact with the world.' Unlike writing and other analogue media that offer both the poison and the cure, the possibility of the expansive microtemporal experience that algorithmic technologies offer 'require[s] a technical supplement *in order to make good on their pharmacological promise*' (50).

Thus, for Hansen we are caught in a double bind: Experience of the world is both opened out in its mediation by contemporary digital media, but at the same time those experiences require new technical supplements as the remedy, which has ethical consequences.

Returning to the example of the iPhone given in Chapter 4, what is perceived as the taking of one photograph is in fact the capturing of multiple images processed into one composite image. Without this additional technical supplement of algorithmic processing the experience of a given moment in time through an image would not make sense. Because algorithmic technologies are able to bypass primary retention, operating on the pre-cognitive sensory continuum, we require machines to register and interpret the sensory data constituting experience. Experience in the infrasomatic city is thus subject to a process of double mediation. Firstly, as experience mediated below the threshold of human perception, and secondly, experience remediated to the level of conscious, cognitive perception. Thus, an object in time bypasses primary retention but is then recycled back through primary retention via a technical support. This captures well the processes at play in infrasomatization that I introduced earlier.

It is the space between these processes of mediation that leaves the experience of everyday life open to manipulation. Hansen and Hayles argue that technologies which operate below the threshold of human perception are operating outside of the grammatization process of inscription and recovery. Grammatization is central to Stiegler's philosophy. Following Derrida and Auroux, he deploys the term to describe the exteriorization of memory in the form of marks, traces or 'the material spatialization of discourse's temporality' (Stiegler, 2010a, p. 145). The question of grammatization in the infrasomatic city is one of the exosomatic exteriorizations of memory in algorithmic technologies that are increasingly making interventions on human experience.

For Stiegler the concept of pharmakon is a means to identify the toxicity of these kinds of technology and also to search for a therapeutics which is an act of resistance—it is a move towards the right to the city. Its usefulness as a concept of resistance is furthered when it is understood alongside another of Stiegler's concepts—neganthropology.

Neganthropology

Neganthropology refers to the dispersal of energy 'posed by exosomatic evolution,' by the constitutive relationship between humans and technology (Stiegler and the Internation Collective, 2021, p. 18–45). Energy may refer to the earth's resources required to power and cool data centres. It may also refer to the expenditure of bodily energy, such as that expended by the attention given to social media companies like Instagram. The definition of energy here is broad. In addition the idea of entropy can be understood (in its thermodynamic definition) as the dissipation of heat and energy from closed technological systems. Physicist Erwin Schrödinger defined entropy as not just the dissipation of energy but the disorganization of the biological, and the destruction of life. Mathematician Claude Shannon defined entropy as the destruction of the value of information. Stiegler defines entropy by using the word 'anthropy.' He uses an 'a' and 'h' to foreground how human life is destroying the earth (Stiegler and the Internation Collective, 2021, p. xxi). To these physical, biological, informational, and anthropomorphic definitions can be added economics, education, politics, social media, or any process where the possibility exists for the disorganization and destruction of everyday life (Stiegler and the Internation Collective, 2021, p. 11).

<u>As Edgar Morin has shown, central to life, which he defines as a struggle against entropy, be that physical, biological, informational, or anthropomorphic, is communication and information (Morin, 1992, p. 300). For Stiegler, this 'is the challenge to find a performative response adequate to all the systemic challenges arising' in our contemporary technological condition (Stiegler, 2018a, p. 31). Stiegler terms it neganthropology, which as defined above can be understood as a methodology capable of thinking the problems 'posed by exosomatic evolution,' that is by the production of bodies by technics (Stiegler and the Internation Collective, 2021, p. 18–45).</u>

> Neganthropology is therefore an 'economy of the *pharmakon*' and as such is an organology. This means that neganthropology seeks to understand the toxic and curative nature of technology by analysing its effects across the biological, social, and technical organs of society, and by understand the entropic or negentropic possibilities of technology (Stiegler, 2019, p. 40).

Stiegler also emphasizes the spatial nature of neganthropology. He states that it is a geography. Which is to say that it is the expenditure of energy in relation to a context and a place (Stiegler, 2018a, p. 45–62). Put simply Stiegler is reminding us that we must think about these matters contextually. That neganthropology is a geography is important for the argument of this book because it means that entropic and negentropic rhythms are always produced in relation to a locality. Therefore, what takes place always has a place and always affects a place and is constitutive of the production of space.

That neganthropology is a geography means that it can also be a useful methodology for understanding what Lefebvre called the right to city.

This is because, as a methodology, it is attentive to shifts in the biological, social, and technical organs of society, to organology. It also seeks the deferral of entropy which could be understood as looking for new ways to design worlds of justice, equity, and hope in the face of the inevitable: The acceleration towards the heat death of the universe.

For example, an area in a city that has been negatively affected by poor infrastructure may cause poor biological health, disconnected communities, and poor social cohesion which—using Stiegler's definition—are the dispersal of energy. Rectifying those infrastructural problems to increase health and social cohesion could be called negantropic. If neganthropology is always a geography, it is a problem of the here and now. And, just as a condition of the possibility of urbanization is alphabetical writing—which is a question of technics—then questions of how we might live towards the negentropic city are questions of grammatology. The extent to which the digital technologies of everyday life produce space, and the extent to which the right to the performance of a negantropic city is possible, are questions of rhythm (Stiegler and the Internation Collective, 2021, p. 8). Neganthropology therefore requires thinking the urban, which is a matter of grammatization. There is, therefore, a history of the urban which is a history of grammatization.

Grammatization

As Stiegler argues, the history of urbanization is a history of grammatization. Urbanization would not have been possible without grammatization. He understands this as the exteriorization of internal human biological drives via external exosomatic technical prosthesis.

Thus, the history of urbanization has always been a history of technological prostheses. The long history of urbanization begins not in the city but in the painted walls of the Upper Palaeolithic caves, the prehistoric acquisition of fire, then the agricultural territorialization of the land, then the invention of alphabetical writing and the exteriorization of laws in written form and now hyper-grammatization through algorithmic digital technologies.

For example, ideographic writing governed the relationship 'between the scribes or clerics or officials of an empire' who serve the Emperor (Stiegler, 2018a, p. 120–121). The advent of alphabetical writing constituted a new form of city which in turn was radically accelerated by the invention of the printing press in Europe. Texts, rules, laws, ideas became easily distributable extending forms of governance across territories (121).

In our contemporary technological condition, grammatization has become digital and increasingly algorithmic and is produced and distributed at speeds faster than human nerve impulses and therefore has the potential to operate on a pre-sensory level including the pre-semiotic (Stiegler, 2018a, p. 121; Rouvroy, 2020). The connection between grammatization and the city is important. Algorithmic grammatization is the production of a new kind of city. Its history is a history of technical systems, which is to say a history of technical exteriorization that spatializes temporal and perceptual flows (Stiegler and the Internation Collective, 2021, p. 20). Therefore, to adequately think the infrasomatic city is to consider the multiple scales across which infrasomatization takes place.

Hayles argues that algorithmic technologies increasingly operate outside of the grammatization process, and that 'control comes in the forms of sensations that precede consciousness and directly address the body's affective responses, leading to the cultural and media phenomena now called affective capitalism' (Hayles, 2017, p. 173). Affective capitalism is the economic system of our contemporary technological condition, where everyday life is constantly being quantified, measured, and processed. It is a system of consumption that produces the very consumers it needs to succeed. Hansen sees the shift from inscription and recovery to pre-conscious sensations as a break in the history of media prosthetics that runs, he argues, from Plato via media theorist Marshall McLuhan to Derrida and Stiegler (173). By inscription, Hansen is referring to the act of writing where something internal to human thought (endosomatic) that is to be later recalled must first be written down externally (exosomatically). By recovery Hansen is referring to the act of reading a written text and that text acting as a memory aid. Algorithmic technologies are fundamentally different because they are able to bypass this process of inscription and recovery by operating below the threshold of human consciousness. Hansen calls this the pre-conscious. Hansen describes this: 'twenty-first-century media directly mediate the causal infrastructure of worldly sensibility... by mediating worldly sensibility, twenty-first-century media simultaneously modulate human sensibility, as it were, beneath the senses' (Hansen, 2015, p. 52). By bypassing the 'slow time resolution of consciousness in order to maximize our material contact with and operational agency over the sensory continuum,' twenty-first-century media transforms temporal regimes into microtemporalities inaccessible to humans (Hayles, 2017, p. 172). While Hansen is right that contemporary digital technology mediates and produces experience in ever more complex ways, I propose that he does not fully understand the nuance of Stiegler's concept of pharmakon. I propose that, contra-Hansen, Stiegler is aware of how different technological epochs produce changes in the pharmakon. This is important for my argument about finding methodologies for architects that are capable of interpreting the multiple levels upon which the toxicity of our age operates.

Stiegler emphasizes that a pharmacology must operate across multiple levels because the toxicity of a given technology is not simply inherent in the technology itself, but operates organologically across the technical, biological, and psychic organs of society.

General Organology

For Stiegler, a general organology is a way to simultaneously think about the technical, social, and the psychic (Hayles, 2017, p. 37). For Stiegler, a general organology must be thought alongside his concept of pharmacology. This is because the toxicity of technology is not simply inherent in the technology on its own, but is connected to the shifting relationships between the psychic, social, and technical. For example, Apple's deep neural network was trained on a dataset of approximately four million images that contained around 1,500 category labels. Additional training was performed using another 50,000 images which were then annotated with (according to Apple) 'extremely high-quality annotations of a handful of categories that can now be predicted completely on-device: sky, person, hair, skin, teeth, and glasses' (CDO 2021). Google, in an advert for the new Google 6 Pro smartphone, promoted the 'Real Tone' feature that can 'accurately photograph all skin tones' (Google, n.d.). This serves as a reminder of the political and ethical dimensions of algorithmic governance and the reality that algorithms 'make the world in their own image' (Amoore, 2024, p. 3).

<u>Thinking these processes organologically is to understand the technical elements at both a hardware and software level (including infrastructures and architectures) required to facilitate the deployment of algorithmic technologies. To think this way is to be immediately confronted with the multiple scales across which these processes operate: Hyperscale data centres, fibre optic cables, tiny processing chips. However, an organological method</u>

means that these technical processes cannot be thought apart from their social implications.

For example, the 50,000 images used in Apple's neural network will have been manually labelled by humans. The crowdsourcing labour software Amazon Mechanical Turk is often used to source this human labour, and the labourers often work in exploitative conditions (Amoore, 2020, p. 72). Using Stiegler's methodology one must think through the psychic implications of algorithmic technologies of this kind. If we only know the world through our representations of it, and those representations are becoming increasingly algorithmically automated to the extent that they operate below the threshold of human cognition, then to what extent does our experience of everyday life change? These psychic implications cannot be thought apart from the social or the technical. Thus, I argue that a pharmacology apart from a general organology is also unable to account for the therapeutic potential of a given technology because the task of a general organology does not just 'describe the toxicity... but prescribe[s] other social arrangements that constitute therapies or therapeutics, that is, systems of care, or attentional forms and of *knowledge*' (Stiegler, 2012, p. 174). Therefore, the question of the extent to which a right to the infrasomatic city is possible is a question of a general organology, which is a question of pharmacology as the prescription of a therapeutics, which ultimately is a question of hope. Any organology must also be a pharmacology.

Multi-Scalar Methodologies

This chapter introduced Stiegler's concept of pharmakon alongside his concepts of organology and neganthropology. The chapter argued that, for architects to become neganthropologists—which is to say to move away from a thinking that is unable to account for the multi-scalar levels upon which our contemporary technological condition operates—will require new methodologies of practice. This requires methodologies and concepts that can account for the complex multi-scalar entanglements produced by our contemporary technological condition.

Using Mark Hansen's critique of pharmakon, I showed that the concept remains useful in the *infrasomatic city*. I argued that Hansen, in his critique of the concept of Stiegler's pharmakon, misses the connection with a general organology. In doing so, he is left unable to see that the toxicity of infrasomatic technologies 'is not inherent in the technology on its own, but derives from those social arrangements brought about by power operating *through* these technologies'—the social, psychic, and technical relations (Stiegler, 2012, p. 174). It is the connection of pharmacology and organology that differentiates Stiegler from media theorists (particularly German) like Fredrick Kittler, Wolfgang Ernst, and others who focus primarily on the historical development and workings and effect of a specific technical object. This is not to say that Hansen's critique is altogether wrong, but I suggest that in failing to connect pharmacology and organology he is left unable to adequately think through a pharmacology of the *infrasomatic city*. This I propose requires revisiting Stiegler's concept of tertiary retention that I introduced in Chapter 3. The next chapter offers an analysis of how Stiegler's tertiary retention might be developed to give an account for how algorithmic technologies operate in the infrasomatic city.

CHAPTER 6

Algorithmic Memory

Introduction

The previous chapter introduced Stiegler's concept of pharmakon alongside his concepts of organology and neganthropology. As an extension of the previous chapter, this chapter proposes how the concept of tertiary retention might be used in our age of algorithmic governance. As noted above for Stiegler, there is a difference between the pharmacological nature of the technological supplement of writing and our epoch of digital technology. Unlike the pharmacological nature of writing, whereby the actual act of writing simultaneously produces both the poison and the cure in our epoch of digital technologies—the curative potential of a particular technology is not given. It is not, unlike writing, inherent in the act of using the technological prosthesis. For example, a word written in ink on a page is the exteriorization of memory in space which is the process of grammartization that I introduced in the previous chapter. For Plato, that something contained within memory is, through the use of an exosomatic prosthesis, made exterior. This means that there is a loss of memory: The substrate of the page becomes a form of memory retention and can allow the writer to delegate the function of endosomatic memory to an exosomatic tool. However, the pharmacological relationship between poison and the cure is made more complex when we consider the example of the iPhone camera or the Rollback Netcode algorithm. With these two examples we saw that between the act of producing a representation of space in the form of image code and the representation of space itself as code image a space is opened out for image manipulation to take place at speeds below the threshold of human consciousness. This book argues that Stiegler's concepts of pharmakon, organology, and neganthropology are the basis of working towards a right to the city. If this is so, then the question of how the concept of pharmakon is able to account for the changes brought about by our contemporary algorithmic age must be considered. What might a pharmacology of technologies that operate below the threshold of human cognition look like?

A Pharmacology of the Algorithm

For Stiegler, all new forms of digital tertiary retention—for example, a smartphone—may be predominantly poisonous unless therapeutics are produced (Stiegler, 2012, p. 28). Stiegler recognizes that, in our digital technological epoch, the production of a therapeutics is increasingly complicated because digital technologies operate at ever increasing speeds and have new levels of access to the body. I suggest that Hansen, in his critique of Stiegler, fails to understand the nuances that Stiegler proposed between the pharmakon of analogue and digital media. The difference between the pharmakon of analogue digital media was explained in the previous chapter using the example of writing and algorithmic technologies that operate below the threshold of human consciousness. Writing requires the process of inscription (writing) and recovery (reading) which is a conscious act. Algorithmic technologies, like the example of the iPhone camera, often operate outside of human consciousness. Indeed, Stiegler is clear: 'It is inevitable that this [digital] *pharmakon* will have toxic effects if new therapies, new therapeutics, are not prescribed' (Stiegler, 2012, p. 28). This prescription, I suggest, is central to finding a right to the infrasomatic city. The possibility of this prescription, which I propose is the possibility of finding methodologies that can understand these complex processes, is now discussed.

The challenge of the digital pharmakon, and this is where Hansen's critique is useful, is that digital technologies are increasingly producing new forms of knowledge and discipline beneath the threshold of human perception. Thus, as Hansen rightly proposes:

> **if writing made up for the loss of interior memory by introducing an exterior or artificial memory surrogate [like writing], twenty-first-century**

media compensate for the loss of conscious mastery over sensibility by introducing machinic access to the data of sensibility that operates less as a surrogate than as a wholly new, properly machinic faculty. (Hansen, 2015, p. 53)

Hansen is suggesting that the increased automation of contemporary digital technology does not just act as an extension of the body. Importantly, for Hansen following Stiegler, the technical supplement of writing 'allows access to experiences that have never been lived by consciousness.' For example, reading, watching, or listening to an account of something that has happened in the past, while not allowing me to experience the event in full, does allow me to have some kind of experience of that event. In contrast, the technologies of the digital pharmakon in the infrasomatic city 'facilitate the feeding-forward of data capturing sensory micro-experiences that not only have never been factually lived by consciousness, but that—because of their microtemporality—fall outside the domain of that which can be lived by consciousness' (Hansen, 2015, p. 37). This, for Hansen, problematizes Stiegler's understanding of how we perceive the world (phenomenology) leading him to argue for a new framework for understanding perception in our age of contemporary digital technology. Stiegler calls this a 'post-phenomenology,' It should he argues comprehend the 'past not lived [and] a future not sensed' without recourse to human consciousness as the primary perceiving agent (Keller, 2021, p. 229; Hansen, 2015, p. 37; Parisi and Goodman, 2011; Warburton, 2017). Thus, I suggest that the possibility of a therapeutics is, for Hansen, a question of the extent to which a technical prosthesis can become a therapeutics. As I proposed above, the extent to which this is possible is problematized in our age of contemporary algorithmic technology. It is problematized because many of the technological processes that mediate our experience of everyday life are hidden out of sight.

Hansen concludes: 'the data made available to consciousness by today's microsensors capture experiences that not only were not, *but can never be*, lived by consciousness' (Hansen, 2015, p. 53). 'What is "new" about twenty-first-century media,' proposes Hansen 'is less their technical disjunction from past media than their opening of the operational present of sensibility to various forms of modulation, including most prominently... capitalist ones' (53). Hansen is less interested in how contemporary digital technologies differ technologically from past media. He is interested in the operational potential of technology. By operational, he refers to the concept of operational images developed by filmmaker Harun Farocki. For Farocki, operational images 'do not represent an object, but rather are part of an operation' (53). Farocki uses the example of military images, such as weapons guidance systems, as an example of imagery whose primary purpose is not to represent a target but rather to perform the operation of guiding a missile towards an object for the purpose of that object's destruction. Artist and geographer Trevor Paglen describes these kinds of images as 'starting to "do" things in the world' (Paglen, 2014).

<u>While all images do things in the world, operational images perform specific tasks like guiding a missile or forming part of a machine learning dataset. Jussi Parikka in his text *Operational Images* (2023a) describes operational images as 'images that primarily operate; ... [that] are not necessarily representational or pictorial' (Parikka, 2023b).</u>

For Hansen, contemporary digital technologies—and the way that they operate in real time—make the present operational. Hansen is interested in how this changes how we perceive the world. He argues that 'twenty-first-century media impact human experience not *in itself* or directly—as has been the case (at least predominately) with media up to now—but *precisely and only insofar as they open up, modulate, and channel the power of worldly sensibility itself*' (Hansen, 2015, p. 226). In short, contemporary digital technologies do not just mediate the world to us, but are actually producing how we sense the world. Hansen's proposal is highly conceptual and, as N. Katherine Hayles comments, it is made with almost no real-world examples. It is therefore difficult, beyond what philosophical credentials the argument may or may not have, to test it as an alternative. However, while critical of Hansen, Hayles does offer three compelling reasons why Hansen's work is important. Firstly, Hansen's argument foregrounds temporality—the present, passing moment—in relation to technical and human cognition (Hayles, 2017, p. 172). Secondly, it recognizes that the technical intervention on human experience is increasingly taking place 'prior to the 100-millisecond range where perception and sensation register for humans,' and that any therapeutics must also take place prior to cognition and thus be technically mediated (172). Thirdly, any intervention must be 'systemic.' Thus a therapeutics is an ontological intervention (172).

The question, then, of a therapeutics, of new forms of knowledge, of new programs which produce new curative rhythmics, becomes a question of the extent to which the right to the infrasomatic city is possible. This is becoming increasingly problematized by the delegation of memory not just to digital tertiary objects, but the delegation of primary retentions to technologies of digital tertiary retention.

This, as Hansen argues, has serious implications for Stiegler's Husserlian-inspired framework of primary, secondary, and tertiary retentions and protentions. Hansen is useful here, in that he proposes that the possibility of a pharmacology remains a question of how a technical supplement may or may not reveal the pre-cognitive levels at which algorithmic technologies increasingly operate. This is now introduced to further develop the argument that any methodology adequate for understanding the complexities of our contemporary technological condition must be able to operate across the same multiple levels upon which the toxicity of our technological age operates.

Algorithmic Tertiary Retention

Our algorithmic age problematizes the process of cognition. 'Mnemonic control,' proposes Hayles, following Hansen and media theorist Shane Denson, 'works on a timeline that operates before... primary retention... and much earlier than [Stiegler's] secondary or tertiary retention' (Hayles, 2017, p. 173).

> It does not depend upon inscription and recovery at a later time through 'grammatological' decoding; rather, control comes in the form of sensations that precede consciousness and directly address the body's affective responses, leading to the cultural and media phenomena now called affective capitalism. (Hayles, 2017, p. 173)

This is significant, as the technologies of our algorithmic age '*precede* the cognitive timeline while also decisively influencing the kinds of sensations and perceptions that are possible and relevant within a given milieu' (173). It is the '*cognitive capacities and their abilities to interact with humans as actors within cognitive assemblages*' that make algorithmic technologies fundamentally different from previous technologies (Hayles, 2017, p. 174).

It is important to note that the unconscious does play a role in Stiegler's framework of primary, secondary, and tertiary retention. For Stiegler, 'the structure of consciousness is thoroughly cinematographic' (Stiegler, 2011, p. 26). Just as cinema is a montage of images, so consciousness is a selection of montages of primary, secondary, and tertiary elements. Our experience of everyday life, our stream of consciousness, is—for Stiegler—like cinema. Consciousness is a montage of primary, secondary, and tertiary retentions. Prior to the ability of tertiary retentional objects to record the present, and playback the past, the selection of primary retentions—that which is experienced as the immediate passing of time—is applied 'solely to secondary retentions of lived, conscious memory' (22).

Secondary retentions, as acts of imagination, influence the selection of the experience of the immediate present. Tertiary retentional technologies of recording, according to Stiegler, have reconfigured secondary retentions' effects on the selection of primary retentions. Recording technologies root primary and secondary retention in one another 'through the technical possibility of the temporal object's repetition which prior to technologies of cinematic consciousness was not possible' (21). For example, that which I have experienced before, and is thus a secondary retention, can, through the repetition of a recording device, be experienced again and again. This is as a primary retention that is both a repetition and difference and a secondary retention that I have already experienced, which therefore means I have an anticipation of what is to come. It is this process that is short-circuited in algorithmic representations of space like those produced by Rollback Netcode.

When I hear a piece of music for the first time, I am unable to anticipate exactly what note will follow the present note. If I am listening to a recording of the same piece of music for the second or third time, the anticipation of the following notes already forms part of my secondary retention. Thus, for Stiegler, tertiary retentional objects—such as a record player—fundamentally change the relationship between primary and secondary retention. Perception becomes a 'montage of overlapping primary, secondary, and tertiary memories.' Making sense of the present becomes the assembly of the montage. Consciousness is the 'post-production centre' of experience assembled by the *unconscious* when the unedited raw footage of experience, the 'rushes and the montage are out of sync' (Stiegler, 2011, p. 39). Algorithmic technologies like Rollback Netcode or the iPhone camera fundamentally change the makeup of these cinematic montages by introducing non-conscious primary retentions, secondary retentions that I may have lived but did not experience, and non-conscious cognitive tertiary retentions.

Contemporary tertiary retention technologies are able to bypass conscious primary retentions and operate on a pre-conscious sensory level. This means that since the rushes and the montages are increasingly out of sync, making sense of experience the post-production centre of consciousness must rely on the algorithm. It is this closing of the loop between the technological tool and the body that Berry captures in the development of the concept 'infrasomatization.'

In this regard, Hansen is right—algorithmic technologies are able to bypass primary retention and operate pre-cognitively, we must delegate the post-production of experience to machines which are able to 'register and interpret the sensory data constituting experience.' It could be suggested that what Stiegler called the cinematic consciousness must now be retheorized as the post-cinematic consciousness, by thinking through the implications of non-conscious retention. The idea of non-conscious tertiary retention is now introduced as a possible advancement of Stiegler's philosophy of memory retention.

Non-conscious Tertiary Retention

While Stiegler did not consider post-cinematic consciousness, he did recognize that industrial temporal objects modify consciousness—although perhaps not to the extent that I have outlined above. He makes a clear distinction between primary retention, which belongs only to the present moment of perception, and secondary retention which is always that which is past. The two must not be confused (Stiegler, 2009a, p. 54), although it could be argued that algorithmic technologies are increasingly confusing this relationship. However, importantly for Stiegler, contemporary algorithmic technologies 'permit in one blow the modification of the process of consciousness' and, it is possible—I would suggest increasingly inevitable—that algorithmic technologies influence and control this process (54). These technologies overdetermine 'the relations between primary and secondary retentions… permitting their *control.*' 'These industrial temporal objects are increasingly today what give rhythm to, and frame, the flux of consciousness that we are' (Stiegler, 2009a, p. 54).

Stiegler emphasizes this further in the—as yet unpublished—fourth volume of *Technics and Time, Faculties and Functions of Noesis in the Post-Truth Age* (forthcoming). Here, building on Chapter 4 of *Automated Society Volume 1, The Future of Work* (2016), he briefly mentions the idea of 'automated protentions' which short circuit the retentional-protentional process. What's more, in line with his proposal in *Acting Out* (2009a) also touching on Chapter 6 of Yuk Hui's *On the Existence of Digital Objects* (2016), Stiegler implies that contemporary algorithmic technologies can modify and control consciousness. He recognizes that retentions and pretentions may be 'pre-conscious' and also 'inaccessible to consciousness: unconscious.' These retentions are *'those of the unconscious'* and '[w]ithin these unconscious retentions, equally unconscious protentions are concealed.' Importantly, for Stiegler, these unconscious retentions and protentions are also *'collectively exosomatic,'* which is to say that they are produced and modulated by collective technologies of synchronization such as those that I discussed above. Thus, I would suggest that if we can talk of collective exosomatization we can also talk of collective infrasomatization. What Stiegler demonstrates, contra Hansen, is that by connecting the concepts of organology and *pharmakon*, toxicity, or indeed the therapeutics of a given technology, is not simply present in the technology. It can only be thought, in context, alongside the always evolving psychic, social, and technical organs. The possibility of a therapeutics, or indeed the toxicity of a given technology, is therefore a question of organology.

What Stiegler also demonstrates is that his retentional-protentional framework does not break down if primary retention is bypassed. Although, as I have suggested above, new terms may need to be introduced into the montage of temporal experience to cater for non-conscious cognitive experience. While he may not have thought through the full implications of this, he does acknowledge as early as 2003 that contemporary algorithmic technologies do in fact modify the process of consciousness.

Philosopher Gilbert Simondon proposed that the cinematographic age would be marked by a 'hypnosis and rhythm that dulls the reflexive faculties of the individual in order to induce a state of aesthetic participation.' In our age of contemporary algorithmic technology, Stiegler understands this participation as leading to a 'symbolic misery,' a malaise produced by a '*loss of participation in the production of symbols*' through ever increasing technological synchronization (Simondon, 2017, p. 116; Stiegler, 2014, p. 10; Stiegler, 2011, p. 120–125).

I suggest that the infrasomatic city is rich in imperceptible rhythms that bypass the primary reflexive faculties of the individual in order to induce a state of passive aesthetic participation. The infrasomatic city re-cognizes everyday life. The prescription of counter rhythms or algorithmic counter governance as therapeutics is a question of the right to the infrasomatic city. While Hayles and Hansen expose the limitations in Stiegler's philosophy of technology for our age of contemporary algorithmic technology, I conclude that we should work with Stiegler's framework of perception and not against it, but do so alongside Hansen's observations about the pre-perceptual nature of algorithmic technologies and within Hayles' framework of cognitive assemblages (Hayles, 2017). Thinking this way is to begin to think the possibility of a right to the infrasomatic city.

A Right to the Infrasomatic City

The speed at which algorithmic image production technologies like GGPO operate means that we now experience non-conscious primary retentions, secondary retentions that we may have lived but did not experience, and non-conscious cognitive tertiary retentions. Algorithmic image production technologies are thus able to operate on a pre-conscious sensory level whereby such consciousness becomes an infrastructural component (what Stiegler would call an organ) of the algorithm. Here, the post-production of experience becomes increasingly delegated to machines which are necessary for us to be able to register and interpret the evermore complex sensory data of everyday life. However, this chapter argues that the answer to the question of the possibility of a therapeutics, of a right to the infrasomatic city, must be considered across the multiple levels on which toxicity operates.

<u>As these processes are briefly exposed we become aware that we have been subject to 'the microtemporal misfirings of the computer into our subjective awareness' (Denson, 2022a, p. 2).</u>

What is produced are 'entirely new configurations and parameters of perception and agency' (21). We, the users, who are also the products, get placed in an unprecedented relation to algorithmic infrastructures. We become the infrastructure.

Algorithmic technologies that operate below the threshold of human consciousness produce a fundamental shift in retention. The possibility now exists for the production of secondary retentions that bypass conscious primary retentions. I conclude that one of the primary technological conditions of the infrasomatic city is the production of imperceptible images that bypass the primary reflexive faculties of the individual. These induce a state of passive aesthetic participation. This leads to an area of future possible research which could explore prescription of counter rhythms or algorithmic counter governance as a therapeutics. This I conclude is the question of the extent to which the right to the infrasomatic city is possible.

This book has argued that the work of Bernard Stiegler is useful for architectural practice. It provides a methodology to enact a right to the contemporary technological city. The question of the extent to which the right to the city is possible in the infrasomatic city is a question of the extent to which it is possible to prescribe a therapeutics that is capable of being a cure across the multiple scales upon which algorithmic technologies operate. To understand how a therapeutics is possible is to understand the changes in the processes of grammartization that have taken place. Grammartization in the infrasomatic city involves the exosomatic exteriorization of memory in algorithmic technologies that increasingly intervene on human experience 'prior to the 100-millisecond range where perception and sensation register for humans' (Denson, 2022a, p. 21). Furthermore, these technologies do not just mediate our experience of the world; they are entangling endosomatic biological organs together with exosomatic prostheses to the extent that the body becomes an integral component of algorithmic infrastructures. This produces endosomatic evolutions such as new neurological flows which in turn produce new ecologies of attention and addiction (Citton, 2017). The question of how it is possible to understand these complex new relationships is a question of finding adequate methodologies capable of understanding the multiple scales across which our contemporary technological condition operates. This book has argued that one such methodology is the work of Bernard Stiegler.

References

91mobiles. (n.d.) 'Deep Fusion: Understanding the Technology Behind Apple iPhone 11's New Camera Capabilities.' www.91mobiles.com/hub/deep-fusion-understanding-the-technology-behind-apple-iphone-11s-new-camera-capabilities (Accessed 5 May 2022).

Aaron, B. (2009) 'IBM Launches a "Smart City" Project in China,' Technology, *Wall Street Journal, Eastern Edition*, September 17.

Amoore, L. (2020) *Cloud Ethics, Algorithms and the Attributes of Ourselves and Others*. Durham: Duke University Press.

Amoore, L. (2024) 'The Deep Border.' *Political Geography*. https://doi.org/10.1016/j.polgeo.2021.102547

Audouze, F. (2002) 'Leroi-Gourhan, a Philosopher of Technique and Evolution.' *Journal of Archaeological Research* 10, no. 4 (December): 277–306.

Back, A. (2009) 'IBM Launches a 'Smart City' Project in China,' Technology, *Wall Street Journal, Eastern Edition*, September 17.Ball, M. (2021) 'Networking and the Metaverse.' www.matthewball.vc/all/networkingmetaverse. (Accessed 30 March 2022).

Barthes, R. (1993) *Camera Lucida*. Translated by Richard Howard. London: Vintage Classics.

Berry, D. (2016) 'Infrasomatization.' http://stunlaw.blogspot.com/2016/12/infrasomatization.html. (Accessed 7 May 2018).

Berry, D. (2018) 'Infrasomatization, the Datanthropocene and the Negantropic University.' https://stunlaw.blogspot.com/2018/09/infrasomatization-and-datanthropocene_22.html (Accessed 22 September 2018).

Bigo, D., Engin, I., and Evelyn, R., eds. (2019) *Data Politics: Worlds, Subjects, Rights*. Oxford: Routledge.

Blom, I., Trond, L., and Eivind, R., eds. (2016) *Memory in Motion Archives, Technology, and the Social*. Amsterdam: Amsterdam University Press.

Broz, M. (2024) 'Selfie statistics FAQs.' https://photutorial.com/selfie-statistics/#:~:text=How%20many%20selfies%20are%20taken,4%25%20of%20which%20are%20selfies. (Accessed 20 June 2024).

Capener, D. (2018) 'In Many Ways Smart Cities Are Really Very Dumb.' www.citymetric.com/fabric/many-ways-smart-cities-are-really-very-dumb-4384

CDO. (2021) 'How Apple Uses AI to Produce Better Photos.' www.cdotrends.com/story/15984/how-apple-uses-ai-produce-better-photos?refresh=auto (Accessed 20 June 2024).

Citton, Y. (2017) *The Ecology of Attention*. Cambridge: Polity Press.

Coral. (n.d.) 'Examples, Code examples and project tutorials to build intelligent devices with Coral.' https://coral.ai/examples (Accessed 4 April 2022).

Dataminr. (n.d.) *Dataminr*. www.dataminr.com/ (Accessed 20 November 2020).

Denson, S. (2022a) *Discorrelated Images*. Durham: Duke University Press.

Denson, S. (2022b) 'AI, Deep Learning, and the Aesthetic Education of the "Smart" Camera.' (Paper presented at SCMS).

During, E., Bernard, S., and Benoît, D. (2017) *Philosophizing by Accident*. Edinburgh: Edinburgh University Press.

Easterling, K. (2016) *Extrastatecraft: The Power of Infrastructure Space*. London: Verso.

Ernst, W. (2017) *The Contemporary Condition, The Delayed Present: Media-Induces Tempor(e)alities & Techno-Traumatic Irritations of 'the Contemporary.'* Berlin: Sternberg Press.

Foucault, M. (2007) *Security, Territory, Population: Lectures at the Collège de France 1977–1978*. Translated by Graham Burchell. New York: Palgrave Macmillan.

Georgescu-Roegen, N. (1970) *The Entropy Law and the Economic Problem*. Oxford: Routledge.

Georgescu-Roegen, N. (2011) 'Energy and Economic Myths,' in M. Bonaiuti (ed.), *From Bioeconomics to Degrowth: Georgescu-Roegen's 'New Economics' in Eight Essays*. London: Routledge Studies in Ecological Economics. 58–92.

GitHub. (n.d.) 'google-coral.' https://github.com/google-coral (Accessed 25 April 2022).

Google. (n.d.) 'The Best of Google Built Around You.' https://store.google.com/product/pixel_6_pro?hl=en-GB (Accessed 2 May 2022).

Greenfield, A. (2017) *Radical Technologies*. London: Verso.

The Guardian. (2022) 'How TikTok's Algorithm Made It a Success: It Pushes the Boundaries'. www.theguardian.com/technology/2022/oct/23/tiktok-rise-algorithm-popularity. (Accessed 21 November 2022).

Halegoua, G. R. (2019) *The Digital City, Media and The Social Production of Place*. New York: New York University Press.

Halpern, O. (2014) *Beautiful Data: A History of Vision and Reason Since 1945*. Durham, NC: Duke University Press.

Hansen, M. (2015) *Feed-Forward: On the Future of Twenty-First-Century Media*. Chicago, IL: The University of Chicago Press.

Harvey, D. (2003) *The New Imperialism*. Oxford: Oxford University Press.

Hayles, N. K. (2017) *Unthought: The Power of the Cognitive Nonconscious*. Chicago, IL: The Chicago University Press.

Hui, Y. (2016) *On the Existence of Digital Objects*. Minneapolis, MN: University of Minnesota Press.

Husserl, E. (2005) *Collected Works, Volume XI: Phantasy, Image Consciousness and Memory (1898–1925)*. Edited by Rudolf Bernet. Translated by John B. Brough. Dorerecht: Springer.

Keller, M., Claus, G., and Florian, A, eds. (2021) *Automated Photography*. Lausanne: ECAL / Morel Books.

Kitchen, R., and Dodge, M. (2011) *Codespace*. Cambridge, MA: MIT Press.

Kittler, F. (2017) 'Real Time Analysis Time Access Manipulation.' *Cultural Politics* 13, no. 1: 1–18.

Lefebvre, H. (1991) *The Production of Space*. Boston, MA: Blackwell.

Lefebvre, H. (2003) *The Urban Revolution*. Minneapolis, MN: University of Minnesota Press.

Lefebvre, H. (2008) *Writings on Cities*. Oxford: Blackwell Publishing.

Lefebvre, H. (2009) *Dialectical Materialism*. Translated by John Sturock. Minneapolis, MN: University of Minnesota Press.

Leroi-Gourhan, A. (1993) *Gesture and Speech*. Translated by Anna Bostock Berger. Cambridge, MA: MIT Press.

Lotka, A. (1925) *Elements of Physical Biology*. California: Williams and Wilkins.

Marcuse, P. (2009) 'From Critical Urban Theory to the Right to the City.' *City*, 13: 185–197.

Mirzoeff, N. (2011) *The Right to Look: A Counterhistory of Visuality*. Durham, NC: Duke University Press.

Morin, E. (1992) *Method: Towards a Study of Humankind: The Nature of Nature*. Translated by J. L. Roland Bélanger. Lausanne: Peter Lang.

Niantic Labs. (2022) 'Build with the Real-World Metaverse.' https://lightship.dev (Accessed 30 April 2022).

O'Halloran, J. (2022) 'Global Broadband Divide Is Closing but Speed Inequalities Widen.' *Computer Weekly*, 25 January. www.computerweekly.com/news/252512375/Global-broadband-divide-is-closing-but-speed-inequalities-are-widening (Accessed 3 May 2022).

Paglen, T. (2014) 'Operational Images.' *e-flux* 59. http://e-flux.com/journal/59/61130/operational-images/ (Accessed 27 May 2018).

Parikka, J. (2023a) *Operational Images*. Minneapolis, MN: University of Minnesota Press.

Parikka, J. (2023b) 'Operational Image.' https://jussiparikka.net/category/operational-image.

Parisi, L., and Goodman, S. (2011) 'Mnemonic Control,' in P. Ticento Clough and C. Willse (eds.), *Beyond Control*. Durham, NC: Duke University Press.

Popper, K. (1972) *Objective Knowledge: An Evolutionary Approach*. Oxford: Oxford University Press.

Pusch, R. (2019) 'Explaining How Fighting Games Use Delay-Based and Rollback Netcode.' https://arstechnica.com/gaming/2019/10/explaining-how-fighting-games-use-delay-based-and-rollback-netcode. (Accessed 10 April 2022).

Roberts, B. (2006) 'Cinema as Mnemotechnics: Bernard Stiegler and the Industrialisation of Memory.' *Angelaki* 11, no. 1: 55–63.

Rouvroy, A. (2020) 'Algorithmic Governmentality and the Death of Politics.' www.greeneuropeanjournal.eu/algorithmic-governmentality-and-the-death-of-politics (Accessed 1 June 2021).

Rouvroy, A., and Berns, T. (2013) 'Algorithmic Governmentality and Prospects of Emancipation.' *Reseaux* 177, no. 1: 163–196.

Rouvroy, A., and Stiegler, B. (2016) 'The Digital Regime of Truth: From the Algorithmic Governmentality to a New Rule of Law.' *La Deleuziana—Online Journal of Philosophy*, 3: 6–29.

Schivelbusch, W. (2014) *The Railway Journey: Trains and Travel in the Nineteenth Century*. Berkeley, CA: University of California Press.

Simondon, G. (2017) *On the Mode of Existence of Technical Objects*. Translated by Cecile Malaspina. Minneapolis, MN: Univocal.

Starosielski, N. (2016) *The Undersea Network*. Durham, NC: Duke University Press.

Stiegler, B. (1998) *Technics and Time: 1. The Fault of Epimetheus*. Translated by Richard Beardsworth and George Collins. Stanford, CA: Stanford University Press.

Stiegler, B. (2009a) *Acting Out*. Translated by David Barison,. Daniel Ross, and Patrick Crogan. Stanford, CA: Stanford University. Press.

Stiegler, B. (2009b) *Technics and Time: 2. Disorientation*. Translated by Stephen Barker. Stanford, CA: Stanford University Press.

Stiegler, B. (2010a) *Taking Care of the Youth and Generations*. Translated by Stephen Barker. Stanford, CA: Stanford University Press, 2010.

Stiegler, B. (2010b) *What Makes Life Worth Living, on Pharmacology*. Translated by Daniel Ross. Cambridge: Polity.

Stiegler, B. (2011) *Technics and Time: 3. Cinematic Time and the Question of Malaise*. Translated by Stephen Barker. Stanford, CA: Stanford University Press.

Stiegler, B. (2012) *Nanjing Lectures 2016–2019*. Edited and translated by Daniel Ross. London: Open Humanities Press.

Stiegler, B. (2014) *Symbolic Misery: 1. The Hyperindustrial Epoch*. Translated by Barnaby Norman. Polity: Cambridge.

Stiegler, B. (2015) *Automated Society Volume 1: The Future of Work*. Translated by Daniel Ross. Cambridge: Polity.

Stiegler, B. (2018a) *The Neganthropocene*. Edited and translated by Daniel Ross. London: Open Humanities Press.

Stiegler, B. (2018b) 'Power, Powerlessness, Thinking, and Future,' *Los Angeles Review of Books*, October 18.

Stiegler, B. (2019) *The Age of Disruption, Technology and Madness in Computational Capitalism*. Translated by Daniel Ross. Cambridge: Polity.

Stiegler, B. (2020) *Nanjing Lectures 2016–2019*. Translated by Daniel Ross. London: Open Humanities Press.

Stiegler, B. (forthcoming) *Technics and Time: 4. Faculties and Functions of Noesis in the Post-Truth Age*. Translated by Daniel Ross.

Stiegler, B., and Derrida, J., eds. (2002) *Echographies of Technology*. Cambridge: Polity Press.

Stiegler, B. and the Internation Collective, eds. (2021) *Bifurcate 'There Is No Alternative.'* London: Open Humanities Press.

Tom's Hardware. (n.d.) 'Apple A15 bionic powers iPhone 13 and iPad Mini.' www.tomshardware.com/uk/news/ipad-iphone-13-a15-bionic (Accessed 5 May 2022).

Warburton, A. (2017) 'Goodbye Uncanny Valley.' Vimeo Video, 14:38. https://vimeo.com/237568588. (Accessed 10 May 2022).

WonderNetwork. (n.d.) 'Global Ping Statistics.' https://wondernetwork.com/pings/Dublin. (Accessed 30 March 2022).

Zuboff, S. (2019) *The Age of Surveillance Capitalism*. London: Profile Books.

Further Reading

My hope is that this book will help start a conversation about how Bernard Stiegler's philosophy of technology can be of relevance to architects in pursuit of a right to the city in our contemporary technological age. At the beginning of this book, I encouraged you to keep walking along the path even if at first read not everything was immediately clear. My hope is that I have laid a clear enough path to give you a solid introduction to Stiegler's often-difficult concepts. I too am still wrestling with his ideas, and will continue to do so in the hope that as I do I will find new ways of thinking about and acting in what at the time of writing is becoming an increasingly precarious world. My invitation to you is to join me. This precarious time in which we find ourselves urgently needs those who are committed to finding new kinds of theories and praxis that open out ways to imagine and design worlds of justice, equity, and hope in the face of the inevitable heat death of the universe.

So, for those who want to continue their journey through Stiegler's philosophy I would suggest the following. Starting with some secondary sources. The introduction and Chapters 1 and 2 of *Stiegler and Technics* (2013) edited by Christina Howells and Gerard Moore will give you a great introduction to Stiegler's philosophy. The book expands on many of the ideas introduced in this book as well as introducing other ideas that could not be included here. Another very accessible text is Chapter 3 of *The New French Philosophy* (2012) by Ian James. This is a highly accessible read and offers a very good overview of Stiegler's foundational ideas. In terms of primary sources, a good place to start is the book *Philosophising by Accident: Interviews with Elie During* (2009). The book is a transcription of four radio interviews that Stiegler gave in 2002, and is probably the best introduction to understanding his philosophy as well as biographical details about his life. Next, *The Nanjing Lectures* (2020) is a free book available online comprising short, concise lectures that he delivered at Nanjing University between 2016 and 2019. The lectures cover most of the major themes of Stiegler's work, and develop the idea of the consequences of the entropic and thermodynamic revolution.

The above books will serve as an excellent grounding for tackling Stiegler's seminal work, the *Technics and Time* series volumes 1–3. Stiegler once told me that the best place to start is volume 3, *Cinematic Time and the Question of Malaise* (2010) and then volume 1, *The Fault of Epimetheus* (1998). Volume 3 will give you a thorough philosophical account of the concept of tertiary retention. Volume 1 investigates the key question that runs throughout Stiegler's work: What is a technical object?

Index

affective capitalism 56–57, 65, 74
algorithmic governmentality 33–36
algorithmic technology, contemporary:
  algorithmic governance 33–34, 36;
  cognitive capabilities, pre-conscious
  influencing 70–73, 75–76, 78–80;
  digital mapping and consumer
  influence 19–20; 'discorrelated
  images,' mediated perceptions
  27–28; infrasomatization 19–22,
  34–39; mediated experiences,
  non-conscious retentions 59–61,
  65–66; mnemotechnical, social
  structuring 39–40; non-conscious
  tertiary retention 76–78; online
  gaming and infrasomatic subjectivity
  22–28; operational images 72;
  passive aesthetic participation 78,
  79; pharmacology of 56–59; political
  and ethical issues 58–59, 66–67;
  proletarianization 40; smartphones
  and tertiary retention 43, 47, 49–52,
  58, 75; Stiegler's right to the city
  concerns 5–6
Amazon Mechanical Turk 67
Amnesty International 37
Amoore, Louise 58

Apple 50, 51, 66–67
Auroux, Sylvain 46

Barthes, Roland 51
Berry, David 19–22, 34–35, 39, 40
blind person's cane 15–16

Cisco 36, 37
Citton, Yves 22
city, the: algorithmic governance 33–34,
  36; Lefebvre's right to city 4;
  Lefebvre's urbanization theory 30–31;
  'smart city,' concept and development
  issues 36–39; 'smart' urbanization
  31–32; Stiegler's 'exorganic milieu'
  32–33
clinic of contribution 5
Codespace 32
consciousness as cinematographic
  74–75

Dataminr 27
Deep Fusion 50, 51
Denson, Shane 27, 78–79
Derrida, Jacques 46
digital mapping and consumers
  19–20, 39

Doctoroff, Dan 37–38
Dodge, Martin 32
During, Elie 48

Easterling, Keller 32
endosomatization: algorithmic
   infrastructure integration 42,
   65, 79–80; conceptual basis
   16–18, 30; grammartization
   65; infrasomatization, concept
   influence 35–36, 40; organology 56;
   tertiary retention and algorithmic
   technologies 42–43, 47, 49–52
entropy 61–62
Ernst, Wolfgang 68
'exorganic milieu,' Stiegler's city 32–33
exosomatization: algorithmic
   infrastructure integration 42, 65,
   79–80; conceptual basis 16–18;
   'exorganic milieu,' Stiegler's city
   32–33; grammartization 61, 65,
   69; infrasomatization, concept
   influence 35–36, 40; organology 56;
   tertiary retention and algorithmic
   technologies 42–43, 47, 49–52
'extrastatecraft' 32

Facebook 20, 47
Farocki, Harun 72
Fitzpatrick, Noel 5
Foucault, Michel 33–34

Georgescu-Roegen, Nicholas 17
Good Game Peace Out (GGPO) 25–28

Google/Alphabet: Google Maps
   19, 38; *Ingress* 38; Sidewalk Labs
   project 36–38; smartphones and AI
   technology 50–51, 66
Google Maps 19, 38
grammartization 46–47, 61, 64–66,
   69, 79
Greenfield, Adam 19

Hansen, Mark: digital pharmakon
   challenges 65–66, 70–73, 76;
   pharmakon, concept critique 56–61, 68
Harvey, David 4–5
Hayles, N. Katherine 56, 61, 65, 72–74
Hui, Yuk 52
human is an invention:
   endosomatization and
   exosomatization 16–18; Stiegler's
   foundational concept 13, 28, 56;
   technological prostheses 14–16
Husserl, Edmund 47–48

IBM 36, 37
infrasomatic city: algorithmic
   governmentality 33–36, 40–41; body
   as constitutive infrastructure 21–22,
   35, 39; 'exorganic milieu,' Stiegler's
   city 32–33; grammartization 64–66;
   neganthropology 61–63; pharmakon,
   critique and argument for 56–61,
   67–68; 'smart city,' concept and
   development issues 36–39
infrasomatization: Berry's concept
   19–20, 34–35; digital technology,

body as constitutive infrastructure 20–22, 35–36, 39, 41, 78–80; digital technology, mediation of experiences 59–61, 65; 'hyper-proletarianization' 40; online gaming and circuits of subjectivity 22–26; 'smart city,' algorithmic infrastructures 37–41

Ingress 38

Instagram 20, 27, 47

Internation Collective 18, 30–31

Kitchen, Rob 32

Kittler, Fredrich 52, 68

latency, online networks: affective capitalism issues 56–57; gaming, geopolitical factors 23–25

Lefebvre, Henri 4–5, 30–31, 63

Leroi-Gourhan, André 17, 44–45

Lightship Augmented Reality Developer Kit (ARDK) 39

Lotka, Alfred 17

Marcuse, Peter 4–5

memory and technology: global spatialization and alteration 52–53, 57–58; grammartization 46–47, 64–66, 69; Husserl's two-part retentions 47–48; memory programs and rhythmics 44–46; primary and secondary retentions 47–48, 74–75; tertiary retention, concept 43, 47–49, 75; tertiary retention, manipulated real-time 49–52, 58, 73–77

Merleau-Ponty, Maurice 15

Mirzoeff, Nicholas 59

mnemonic control 73–74, 76–77

mnemotechnical, social structuring 39–40

Moore, Gerald 5

Morin, Edgar 62

neganthropology 61–63, 67

Niantic Labs 38, 39

online gaming, circuits of subjectivity: GGPO's restimulated representations 25–28; latency, geopolitical issues 23–25; network instability 22–23

operational images 72

organology: conceptual basis 13–14, 28; human is an invention 56;

pharmacology, general approach 66–67

Paglem, Trevor 72

Parikka, Jussi 72

passive aesthetic participation 78, 79

pharmacology: concept appliance 5; conceptual basis 14, 54; general organological approach 66–68

pharmakon: algorithmic technology, toxicity potential 56–59; conceptual basis 28–29; digital technologies, Hansen's challenges 65–66, 70–72, 76; infrasomatic city, critique and argument for 56–61, 67–68; prosthesis, poison and cure 55

Plato 55, 58, 69

Pokémon GO 38
Popper, Karl 17
post-phenomenology 71
programmatology 44–46
proletarianization 40

railways, spatial transformers 18–19
Real Smart Cities, The 5–6
rhythmics and memory programs 44–46
right to the city: negentropic transformations 62–63; Stiegler's focus 5–6
Rollback Netcode 25–26, 60, 69, 75
Rouvroy, Antoinette 33–34, 36

Schivelbusch, Wolfgang 18–19
Schrödinger, Erwin 62
Shannon, Claude 62
Sidewalk Labs 36–38
Siemens 37
Simondon, Gilbert 33, 77
'smart city': algorithmic infrastructures and infrasomatization 37–41; concept issues 36–37
smartphone: camera images, manipulated real-time 49–52, 56–58, 75; memory repository 43, 47; political and ethical issues 66–67; technological prosthesis 14–15
Stiegler, Bernard: biography 1–2; Real Smart Cities project 5–6

subjectivity, circuit of: endosomatization and exosomatization 16–18; infrasomatization 20–22, 39; online gaming 22–26; proletarianization 40; technological prosthesis's entanglement 15–16
'symbolic misery' 77

technological prostheses 14–16
tertiary retention: algorithmic 73–76; conceptual basis 47–48; consciousness as cinematographic 74–75; digital pharmakon, mediation of experiences 70–71; iPhone camera, memory repository 43, 47; memory, technical constitution 48–49; non-conscious 76–78; smartphone images, manipulated real-time 49–52, 75; spatial properties 43, 47

urbanization: grammartization's significance 64–65; Lefebvre's process and scales of 30–31; spatial infrastructural technologies 31–32

writing, act and forms: digital media, impact comparisons 58–60, 69, 70; pharmakon, poison and cure 55, 69; urbanization, development role 63–64

X (formerly Twitter) 20, 27, 47